QUARKS AND LEPTONS: THE NEW ELEMENTARY PARTICLES?

QUARKS AND LEPTONS: THE NEW ELEMENTARY PARTICLES?

PROCEEDINGS OF
A ROYAL SOCIETY DISCUSSION MEETING
HELD ON 14 JUNE 1985

ORGANIZED AND EDITED BY
R. H. DALITZ, F.R.S., AND P. I. P. KALMUS

LONDON
THE ROYAL SOCIETY
1986

Printed in Great Britain for the Royal Society
by the
University Press, Cambridge

ISBN 0 85403 276 2

First published in *Proceedings of the Royal Society of London*,
series A, volume 404 (no. 1827), pages 149–298

Published by the Royal Society
6 Carlton House Terrace, London SW1Y 5AG

CONTENTS

Proc. R. Soc. Lond. A **404**, 151–152 (1986)
Printed in Great Britain

Preface

By R. H. Dalitz, F.R.S., and P. I. P. Kalmus

Our concept of elementarity in Nature has changed as we have been able to probe more and more deeply into matter. This was accomplished by the use of particles with ever greater energies and hence smaller quantum mechanical wavelengths. First molecules and atoms were discovered, then nuclei, then protons and neutrons. When energies became comparable with the masses of the particles being probed, new particles were created in the interaction. Since about 1950 a vast spectrum of hadrons (mesons and baryons) has been found, comprising more than 200 named particles.

Over the last decade there has been a significant simplification. Our understanding of physical phenomena at high energies has come to require only two types of particles (quarks and leptons), together with the quanta (photons, gluons, W and Z bosons, gravitons) whose exchange between these particles carries the forces that we observe to act between them.

When the proton–antiproton collider at CERN began to operate in 1981, it quickly became apparent that collison processes involving large energy and momentum transfer have a rather direct interpretation in terms of basic quark and lepton interactions. In consequence, the area of most significant experimentation has moved from the study of interactions between hadronic particles to a direct study of interactions involving quarks, leptons and the force-field carriers.

The recent discovery of the W and Z bosons represented a major step forward in establishing the unity between electromagnetic and weak interactions, a unification of natural forces as fundamental as and in the same spirit as Maxwell's unification of electric and magnetic interactions and light at the end of the last century. Currently, attempts are being made to incorporate also the remaining forces (the strong interactions and gravity) into a unified theory, but achieving this appears to require additional particles, or deeper levels of structure, for which we do not yet have experimental evidence.

A Discussion Meeting was held at the Royal Society on 14 June 1985. It was aimed to be expository in style, up to date, but sufficiently elementary to be understood by scientists outside the field of particle physics. It was attended by about one hundred scientists of whom over half were not particle physicists.

The first talk, outlining the present status of quarks and leptons, was given by Dr Frank Close. He reviewed the excitation of composite systems, which by means of scattering experiments has led us from molecules to quarks and leptons. Free quarks have not been seen, and are now not expected to exist. The most direct manifestation of quarks and gluons is their appearance as jets of conventional hadronic particles in high-energy collisions, and this topic was reviewed by Dr Robin Marshall. He showed how jets were expected to occur, and the extent to which each can be classified in terms of its originating entity (gluon or type of quark), how jet signals can be found in collider experiments, and how they can be interpreted in terms of the quark–gluon interactions which give rise to each observed event.

The W and Z particles were discovered in 1983 at the CERN proton–antiproton collider, and in the following two years many of their properties have been measured. The agreement of these properties with predictions of the so-called 'standard model' is impressive, and this work was reviewed by Dr Allan Clark. Some featuies of the experiments may, however, not be explicable within this model. The evidence for these was discussed by Professor John Dowell, who also presented the status of the evidence for the existence of the top quark, the sixth and possibly final quark in terms of increasing mass.

Historically, the main probes for elucidating the structure of the proton and neutron, have been leptons. The interactions between leptons and hadrons were covered by Dr David Saxon who discussed the similarities and differences observed using electrons, muons and neutrinos. The sensitivity of lepton probes, and the possibility of new phenomena will be increased greatly in the future when the first lepton–hadron collider will start to operate.

A small number of important experiments in particle physics do not use beams from accelerators. Some very large experiments are sited in deep underground locations, to be shielded from the bulk of cosmic radiation. Their primary aim is to search for decays of protons, but they are also sensitive to some astrophysical phenomena. This topic was covered by Dr Tegid Jones.

Although the standard model of the electroweak interactions has been extremely successful, more complete theories are required to give answers to deeper questions, such as the origin of particle masses, and whether the remaining forces – the strong interaction and gravity – can be unified with the electroweak force to form a grand unified theory. The various approaches to these problems were outlined by Dr John Ellis. Finally the likely possibilities, together with some of the more speculative options, for the experimental study of particle physics in the future were discussed by Professor Peter Kalmus.

Proc. R. Soc. Lond. A **404**, 153–166 (1986)
Printed in Great Britain

The present status of quarks and leptons

By F. E. Close
Rutherford Appleton Laboratory, Chilton, Didcot, Oxfordshire OX11 0QX, U.K.

Historically the search for the fundamental elements of matter has exhibited a recurring theme. Patterns are discerned that hint at substructure in the elements, and then scattering reveals the new layer of elementarity. A force is identified that binds the new constituents to build up the old 'elements' as compounds. The Pauli principle is crucial in limiting the possible structures. This is familiar for the atomic elements; it is also true for the strongly interacting 'elementary particles' built from quarks.

These avenues are examined for hadrons, and evidence for quarks, gluons, and colour forces will be reviewed. The theme illustrates how the arena of elementary particle physics has progressed from hadrons to quarks and leptons.

1. The excitation of composite systems: from molecules to quarks and leptons

To see things you must bounce radiation off them. Visible light can only resolve structures larger than about 10^{-3} mm; to probe the make-up of molecules, atoms and nuclear particles requires beams whose wavelength is less than that of visible light. The complementarity between distance and momentum implies that short wavelengths are attained only at high momenta. Thus the ability to probe particles at short distances has improved in parallel with the development of high-energy particle accelerators.

As a result we have uncovered rich and detailed structures within matter and the answer to the question 'What are the fundamental constituents of matter?' has developed over the years. At the start of the century, atoms were the candidates. Within a few years the answer had moved on to 'electrons and positively charged atomic nuclei', by 1932 it was 'electrons, protons and neutrons', and today it is 'leptons and quarks'. The electron is one example of a family of particles known as leptons; protons and neutrons (and other short-lived particles responding to the strong nuclear forces) are all built from quarks.

It is immediately apparent that throughout all these developments the electron has remained on the list. The route from nuclei to protons and neutrons and then to the multifarious hadrons in cosmic rays and man-made accelerators has taken us through a confusing proliferation of particles. Often it may have appeared to the outsider that high-energy physicists were merely collecting new particles as if they were flora and fauna to be collated. It is as a result of these long and painstaking investigations that a new simplicity has at last emerged with the identification of the quark layer of matter.

The discovery of the reality of quarks is much more than the identification of a new layer of matter. It is becoming apparent that at this level, Nature's code becomes rather clear. Indeed, the quarks are in several ways very similar to the electron and leptons. A sense of order has suddenly emerged.

This is surely no accident and promises to be most profound. Nuclear matter appears to be utterly unrelated to electrons. Even at the proton and neutron layer the disguise is almost impenetrable. Only after Nature's packages have been unwrapped down to the quark layer does the uniformity become visible. This promises to be a great breakthrough and was utterly unexpected even 15 years ago, as the quarks first began to emerge from the then still hazy images of protons and neutrons.

Structured systems have the common feature that the constituents can be rearranged relative to one another. Quantum mechanics limits these states to a discrete set, each with a characteristic energy, and as a result the system can exhibit a discrete spectrum of energy levels. Familiar examples include molecules, where the constituent atoms vibrate against each other; atoms, where the electrons can be excited into various orbits; and atomic nuclei whose neutrons and protons vibrate and orbit in various configurations.

The smaller the system, so the greater is the scale of excitation energy. For atoms the excitations are of the order of an electronvolt. For molecules the scale is a fraction of this, whereas for nuclei the scale is a factor of a million times greater.

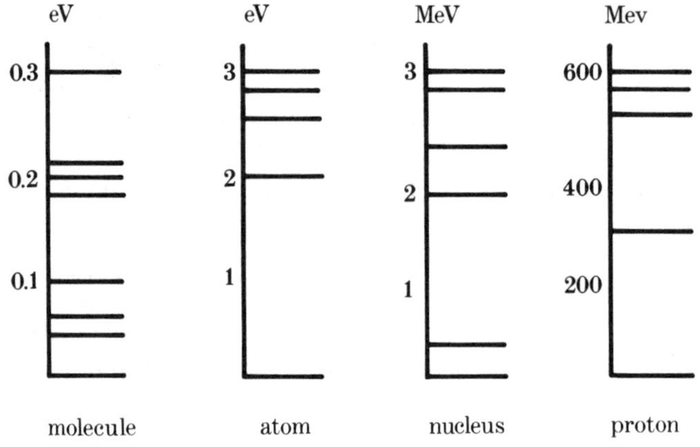

FIGURE 1. Comparative excitation spectra for molecules, atoms, nuclei, and hadrons.

During the 1960s similar excitation spectra were identified for the hadrons (strongly interacting particles, like neutrons and protons). When a proton is hit by a beam of particles, it may be excited into a short-lived resonance state, much as atoms, molecules, or nuclei are. The only essential difference is one of scale: the proton excitations involve tens or hundreds of megaelectronvolts (Wilkinson 1981).

The reason for these excitations is much the same as before. The proton is a composite system built from quarks and it is the different quantum states adopted by the quarks that generate the resonance states.

The proton is the lightest example of a system formed by three quarks; the antiproton is the corresponding system built from three antiquarks. A single quark and a single antiquark form the 'meson' family of hadrons whose lightest example (quark and antiquark in the ground state) is the pion. No other combinations of one, two, or three quarks and antiquarks have ever been seen (Close 1979).

In all composite systems, the excited states tend to be short-lived, decaying to the stable ground state. For molecules and atoms we can literally see the evidence all around us. The carbon nucleus contains an excited level that is in fortunate near-coincidence to the collision energy of three α-particles. As a result, the chance of three α-particles fusing in a star is not small and so the heavy elements essential to life are seeded. The carbon nucleus is formed in an excited state and then emits energy and falls to the stable ground state. Similarly the hadron resonances are short-lived and unfamiliar in everyday life. They decay to ground-state configurations. For systems of three-quarks it is the proton or the neutron (two different varieties of quark, up and down, generate these two independent combinations), which are the stuff of atomic nuclei. Similarly, a system of quark and antiquark will form a pion but only for so long as the quark and antiquark do not mutually annihilate: the pion is unstable. Depending upon the up- and down-quark (and antiquark) content in the original pion, the decay products consist of photons, or of electron or positron accompanied by antineutrino or neutrino.

Leptons and quarks are the most fundamental particles of matter that we have identified so far. However, it may be that there is a layer of reality beyond leptons and quarks, entities which cluster to build them as composite structures, and of which we are currently unaware through lack of powerful instruments.

Whether or not this is so, there is no doubt that quarks and leptons are an essential layer of reality. Their behaviour governs the properties of bulk matter. There is an elegance in the way Nature operates at the quark level which is often hidden at the atomic, let alone the macroscopic, domain. So, irrespective of what may lie beyond, it is important to understand the behaviour of quarks for they represent the deepest level at which we have yet managed to probe Nature's secrets.

This will form the thrust of part of my talk and many of the following speakers. I will briefly survey the issues concerning the fundamentality of leptons and quarks by reviewing the route that led us to these particles and seeing what guides history may give. Many details will be filled in or developed by other contributors at this symposium.

2. Patterns, scattering and structure

A recurring theme has been present during the march to the fundamental constituents of matter (Close 1983). At any given moment in history there are candidates for '*the*' basic elementary objects. Initially these are few in number, and there is optimism that simplicity and order has arrived. There is the promise that the many combinations of these few elements will generate the richness and variety previously discerned; diversity is equated with complexity and lack of fundamentality.

However, this simplicity is a myth. As we probe the elements in more detail we find that there are many more of them than at first we thought. The psychological belief that *the* basic elements are few in number makes one suspect that these 'elements' are not the end of the road. This suspicion becomes scientific when we discover that the elements are not all independent, that several of them appear to be systematically related in their properties. If they were all truly independent there should be no obvious systematic relations among them. This sequence of events is familiar for the atomic elements and I will dwell on it briefly so that some parallels with recent events in elementary particle physics may be drawn.

The first step is that the regularities among the elements are quantified and listed according to their shared properties. This reveals a regularity in properties, Mendeleev's Periodic Table of the atomic elements. I will refer to this first step as identifying *a pattern*.

Gaps in the pattern enable one to predict the existence of other 'elements', which are subsequently found experimentally. Predicting the existence of new forms of 'fundamental' matter from the behaviour of other forms confirms that the scheme is correct and suggests that the elements are not basic at all. The confirmation of the pattern has proven to be the first hint of a deeper layer of reality: The elements are in fact composite structures made from common constituents. In Mendeleev's work these are the electrons and atomic nuclei (protons and neutrons).

The next ingredient is to make direct observation of the new layer of matter. This involves *scattering* high-energy beams of radiation particles, from a target built of the hypothetical constituents. 'Seeing' involves scattering radiation, such as light, from the object of interest and recording the scattering in a detector, such as an eye. To resolve small objects requires beams whose wavelengths are smaller or comparable with the composite structure. Short wavelength corresponds to high energy, and so it is high-energy scattering that resolves how matter behaves at these extremely small distances.

The existence of the deeper layer is finally proved beyond doubt when the constituents are produced as individuals in the laboratory: atoms are ionized and free electrons and nuclear particles obtained. These are available for study in their own right and have unique properties that make them available for specialized technological and medical exploitation.

By the late 1930s it was clear that atomic nuclei consist of neutrons and protons held together by a strong force. To understand the nature of this force, nuclear physicists adopted the strategy of bombarding nuclei with protons and neutrons to see how they responded. High-energy beams probed the nucleus at short distances. Not only were nuclear excitations seen, but a vast spectrum of ephemeral hadrons (strongly interacting particles) was discerned, and since 1950 has grown into more than 200 named particles.

Many of these are excitations (Dalitz 1966): the proton and neutron are merely the lightest of one class of hadrons (known as the baryons) and in consequence the most stable. However, several different varieties were seen, each with their own characteristic excitation spectrum. A pattern (the 'Eightfold way') was discerned, relating the properties of many of these particles and, like Mendeleev, gaps were noted (Gell-Mann & Ne'eman 1962). New particles were predicted to fill the gaps and within a year or two they had been discovered in the laboratory. Here again the pattern is due to a deeper layer of reality: all of the hadrons are built from 'quarks', particles which have some tantalizing similarities to leptons. And it is the various energy configurations available for the quarks that generate the hadronic excitations.

The closest analogies to the α-particle scattering which revealed the nucleus have been experiments which scattered high-energy beams of leptons from nuclear targets and resolved the individual quarks with the neutrons and protons (D. H. Saxon, this symposium). However, unlike the atomic example, where the constituents can be liberated, the quarks appear to be permanently confined in clustered groups: the hadrons. Isolated free quarks have not been produced (R. Marshall, this symposium). Nevertheless we have no doubt that quarks are an essential layer of reality, as subsequent talks will elaborate.

Having identified a new layer of matter there are important questions to ask. What laws govern the behaviour of the new elementary particles? What forces cluster them together to build up the complex structures that had been previously thought of as the elements? Can previously unexplained phenomena at the structured level be understood in terms of the new constituents?

There is an important quantum mechanical principle, the Pauli exclusion principal, that forbids more than one spin-$\frac{1}{2}$ particle to be in the same quantum state. Thus in atoms, one electron can spin up, one down, and that is all that are allowed in the ground state. This underwrites the closed shells in the electronic configurations of the atomic elements. The quarks have spin-$\frac{1}{2}$ and so obey the same exclusion principle. Yet three identical quarks are found to occupy the ground state when forming some hadrons like the proton. This paradox was resolved when the quarks were found to possess a further property, a new form of charge known as 'colour', which distinguished among the three quarks in the hadron and so avoided the exclusion principle.

In the atomic example the force had been known for centuries. The attraction of opposite electrical charges gives us electrostatics. When combined with quantum theory and relativity, the resulting theory of the electromagnetic field and the interactions between electromagnetic radiation and matter is quantum electrodynamics, QED.

There are many subtle effects in QED manifested in the behaviour and properties of atoms. There are tantalizing parallels between these and the behaviour of quarks.

Quarks carry electrical charge and so feel the electromagnetic force. In addition they possess a new form of charge called 'colour'. This acts in some ways like electrical charge but with some subtle and far-reaching differences. As electrostatics and electromagnetism are to electrical charge, so are chromostatics and chromomagnetism to colour.

It is the colour charges on the quarks which attract them to one another. Unlike charges (colours) attract, and three quarks, each charged with a different colour, can mutually attract. A fourth quark will be repelled by the like colour that is already present because there are only three colours available. Hence the colour attractions saturate after three different coloured quarks have attracted and so the neutron and proton naturally emerge as clusters of three quarks.

There are some analogies here to the attraction between positive and negative electrical charges in atoms. The charges neutralize and saturate the force. Molecules are built from the residual effects of the underlying electrically charged particles and forces. By analogy the strong nuclear forces can be thought of as the residual effects of the powerful colour forces at work within the proton and neutron. This is an area of active study for nuclear theorists who are attempting to gain more fundamental insights into the nuclear forces by focusing attention on the colour forces acting among the various quarks.

Chromostatics is similar to electrostatics. Combined with relativity and the quantum theory we have quantum chromodynamics (QCD), by analogy with quantum electrodynamics. This theory of quark interactions has survived a decade of intense scrutiny so far and will be discussed by other speakers here.

3. Quarks and leptons

As far as we know, all matter is made of quarks and leptons. They all have the same intrinsic angular momentum or 'spin', namely 'spin-$\frac{1}{2}$', or half the basic quantum mechanical unit of angular momentum – the smallest allowed quantum of spin.

There are six kinds of quarks, named 'up', 'down'; 'charmed', 'strange'; 'top', 'bottom'. They are denoted by the letters u,d,c,s,t,b. Each quark can carry any of three varieties of the colour charge. There are also six varieties of lepton: colourless particles such as the electron which do not exert colour forces and do not respond to the strong nuclear force. In addition to the (well known) electron there is the neutrino, which always accompanies the electron in radioactive β-decay processes. There are two heavier relatives of the negatively charged electron, each of which is negatively charged: these are the muon and the tau. In turn these each have their own variety of neutrino. This gives a total of six species of lepton, which are denoted by the letters e, ν_e (electron and its neutrino); μ, ν_μ (muon and its neutrino); τ, ν_τ (tau and its neutrino).

The fundamental electromagnetic, colour, and weak forces act on quarks and leptons. Electromagnetic forces act on the electrical charges; the colour or strong force acts on the colour charges and the weak force has its most familiar manifestation in radioactive β-decays.

The way that quarks and leptons respond to these forces specifies the properties of these particles.

If we define the proton charge as our basic unit of electric charge then there are three leptons with an electric charge of -1: the electron, muon and tau. In addition there are three neutrinos that have no electrical charge at all. There are three quarks that have electrical charge $\frac{2}{3}$: (up, charm, and top) and three with electric charge $-\frac{1}{3}$ (down, strange, and bottom), all given in table 1.

Table 1. The fundamental particles of matter

quarks	charge $= \frac{2}{3}$	u	c	t
	charge $= -\frac{1}{3}$	d	s	b
leptons	charge $= 0$	ν_e	ν_μ	ν_τ
	charge $= -1$	e	μ	τ
generation number		1	2	3

We believe that the world is built from up and down quarks, together with electrons and their neutrinos. This set of particles have all the variety known: particles with and without electrical charge and with and without colour. Yet Nature has repeated this pattern (at least) twice. Each of these sets is known as a 'generation' of quarks and leptons. Thus we know of three generations of fundamental particles of matter.

In any generation there are quarks and leptons whose properties are identical to those in another generation except for their masses. For example, the charmed quark is heavier than the up, while the top is heavier than the charmed variety, but otherwise these quarks seem to respond in precisely the same way to electromagnetic, colour and weak forces. Similar remarks apply to the three quarks with charge $-\frac{1}{3}$; the strange quark is heavier than the down and the bottom is an order of magnitude heavier than the strange, while in other respects these quarks appear to act in a similar manner.

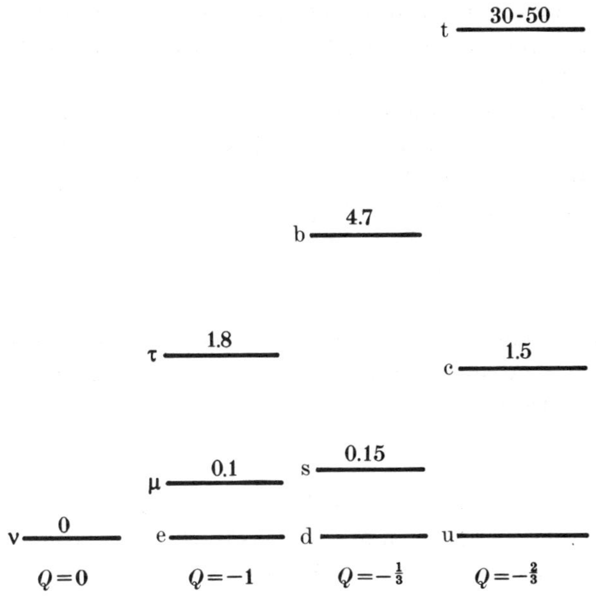

FIGURE 2. Representation of the spectrum of quarks and leptons: Q denotes electrical charge, masses are in gigaelectronvolts and symbols are given in table 1.

For each type of quark or lepton there also exists an antiparticle. The mass of an antiparticle is identical to that of the corresponding particle; its electrical and colour charges are opposite in sign but equal in amount to that carried by its particle counterpart. Thus if by some convention we define the quarks to have positive colour charge, then the antiquark has negative colour charge. Thus the up, charmed, and top quarks (u, c, t) with electrical charge $\frac{2}{3}$ have antiparticles, denoted $\bar{u}, \bar{c}, \bar{t}$, with electrical charge $-\frac{2}{3}$. The $\bar{d}, \bar{s}, \bar{b}$ each have charge $+\frac{1}{3}$ (d, s, b, quarks having charges $-\frac{1}{3}$). Finally the positron (anti-electron) has charge $+1$ as do the positive muon and the positive tau.

(a) The first generation of quarks and leptons

The first generation, consisting of up and down quarks, the electron and neutrino, appears to have all the variety needed to construct our familiar surroundings. The up and down quarks build protons and neutrons. Two ups (each with $+\frac{2}{3}$ electrical charge) and one down (with $-\frac{1}{3}$) combine to give the singly charged positive proton, whereas two down and one up gives the uncharged neutron. Although the neutron has no net electrical charge, it contains charge within it, carried by its constituent quarks. Thus the neutron has a non-vanishing magnetic moment.

The masses or rest energies of the atomic templates are as follows: neutron, 939.573 MeV; proton, 938.280 MeV; electron, 0.511003 MeV; neutrino, less than 0.000046 MeV (Lubimov 1980). Beta-decay of the neutron is an essential ingredient in the way that the Universe operates. In particular it plays a central role in forming the heavy elements necessary for life. Note the delicacy of the masses: the neutron can decay into proton, electron, and neutrino because it is more massive than those three particles combined. We still have no detailed explanations of these important values.

There are three known varieties of neutrino. There has been a claim that the lightest has a mass greater than 0.000020 MeV (20 eV) (Lubimov 1983), but this has not yet been independently confirmed. It is a natural prejudice to regard the neutrinos as massless. The upper limits are as follows (actually allowing quite large masses): $\nu_e < 0.00046$ MeV (more than 0.000020?) (Lubimov 1983); $\nu_\mu < 0.50$ MeV; $\nu_\tau < 143$ MeV (Matteuzzi 1984). So the second and third neutrinos respectively could be as massive as the electron and exceed the muon. Even the mass of the lightest neutrino could be important in cosmological questions. The number density of neutrinos in the Universe is on average more than 100 million times greater than that of protons,

$$N_\nu/N_p = 10^{9\pm1}.$$

So if the neutrino mass is of order 10 eV, which is 10^{-8} times that of the proton, the neutrinos could contribute as much mass to the Universe as do protons.

No *stable* neutrino can have a mass between *ca.* 0.0001 and 2000 MeV without contributing to the energy density of the Universe in excess of observational limits (Gershtein & Zel'dovitch 1966; Cowsik & McClelland 1972; Lee & Weinberg 1977; Vysotsky *et al.* 1977). *Unstable* neutrinos (Dicus *et al.* 1977), could undergo radiative decay by the intermediary of virtual charged particles. The isotropy of the background radiation and the survival of primordial deuterium place some constraints. For example, the τ-neutrino (ν_τ) must be lighter than 1 MeV (Cooper & Sarkar 1984), much below the present empirical limit on its mass.

The masses of the up and down quarks are not easily defined because they are small and the quarks are always confined in clusters. The inferences are indirect and involve the scattering of high-energy leptons from protons, under kinematic conditions where the quarks can be resolved within the target. It appears that these quarks have masses of a few mega-electronvolts at most.

The mass of the proton arises from the kinetic energy of the relativistic confined quarks. The quarks behave as if they are almost free particles within their confinement sphere of some 10^{-15} m. The confining forces seem to act rather abruptly as the quarks separate from one another. This phenomenon has given rise to bag models, where the quarks act as though they are free inside a sphere and the sphere is treated as an infinitely high potential barrier, (Chodos *et al.* 1974; Close 1979).

A massless quark confined in such a volume gains an energy of some 360 MeV. Three quarks therefore give 1080 MeV.

Within atoms there are important spin-dependent forces such as the magnetically induced hyperfine splitting of the ground-state energy of hydrogen. If the electron and proton spin parallel (triplet), the energy is raised; if they are spinning antiparallel (singlet), the energy is lowered. An analogous phenomenon occurs when colour forces act. The colour magnetism separates the energies according to the overall spin state of the three quarks. If they are parallel, spin-$\frac{3}{2}$, the energy is raised and the resonance $\Delta(1232)$ results. If they are in the overall spin-$\frac{1}{2}$ configuration the energy is correspondingly lowered and the proton at 938 MeV results. The 300 MeV photon emitted in transitions between $\Delta(1232)$ and the proton is thus analogous to the 21 cm line in the hydrogen atom.

(b) The spectrum of quarks and leptons

As we have discussed, the three quarks of a given charge or the three leptons of a given charge appear to be identical in all properties but for their masses. This suggests that the heavier members may be excitations of the ground state and that here we have the first hints of a structured system with its associated energy levels.

This first impression may be incorrect. Their behaviour and properties do not fit in with the naïve interpretation.

Figure 2 shows the quarks and leptons masses in giga-electronvolts, the first generation being set at zero mass. Qualitatively this looks similar to the excitation spectra that we have seen for molecules, atoms, and nuclear systems. However, in all of those examples the excited states decay to the ground states by radiating photons, whereas radiative transitions among the quarks or the leptons are not observed.

The most stringent limit is that on the transition

$$\mu^- \rightarrow e^- \gamma.$$

If this occurs at all, it is at most 1.7×10^{-10} of all muon decays (Kinnison *et al.* 1982). For the τ the limit is less good. The transitions $\tau \rightarrow \mu\gamma$ and $e\gamma$ are less than 12% of all decays (Burmester 1977). Similarly, in electron–positron annihilation, we find particle and antiparticle produced pairwise. A negative and positive muon always partner one another, never a muon and an electron. The same appears true for the quarks, though the limits here are not so severe as in the muon decay above. (Radiative transitions do occur for the quarks but only in conjunction with virtual weak interactions: see figure 3). Thus $e^+e^- \rightarrow \gamma \rightarrow \mu^+\mu^-$; $\tau^+\tau^-$; $u\bar{u}$; $d\bar{d}$; $s\bar{s}$; $c\bar{c}$; $b\bar{b}$. At LEP we look forward to seeing top quarks and top antiquarks being produced together.

Some property of matter supresses electromagnetic radiative transitions among

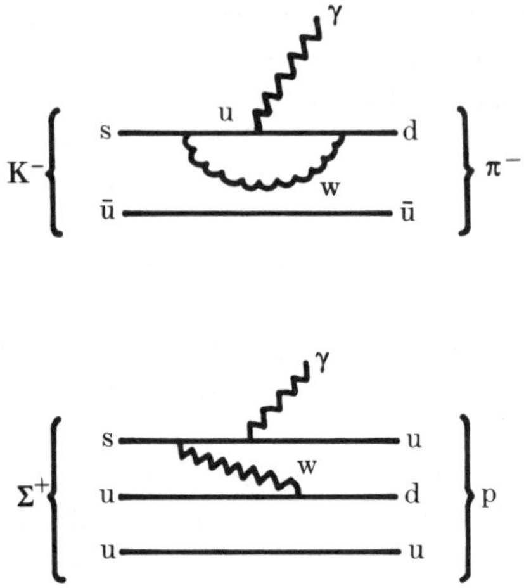

FIGURE 3. Weak flavour-changing transitions conspire to yield an effective flavour-changing radiative transition, as in $K^- \rightarrow \pi^- \gamma \rightarrow \pi^- e^+ e^-$ and $\Sigma^+ \rightarrow p\gamma$.

different varieties. If you wish to build composite models of quarks and leptons this supression is a powerful constraint. In is *prima facie* evidence that the quarks and the lepton are more subtle than 'mere' excitations of the ground-state lightest members.

The heavier leptons or quarks do decay and return to the ground states but only by virtue of the weak interaction. The two families of quarks have electrical charges differing by one unit; so do the two sets of leptons. Transitions occur from one set of quarks (leptons) to the other set of quarks (leptons) through the intermediary of a charged W boson. Moreover, the properties of the interactions among the quarks and the W appear to be identical to those of the leptons and the W. As far as the β-decay is concerned the leptons and quarks are indistinguishable.

This is the first example of the similarity between the leptons and the quarks referred to in §1. The β-decay of the neutron shows only a marginal resemblance to that of the muon for example. However, they are identical at the level of quarks. A single down quark in the neutron turns into an up quark, following the same behaviour as when leptons undergo β-decay, Changing a down quark into an up quark also changes a neutron into a proton, $N(ddu) \rightarrow P(udu) + e^- + \bar{\nu}_e$. It is the correlations of ups, downs, and spins within the neutron and proton that obscure the fundamental Fermi–Gamow–Teller structure at quark level and gives complicated relations among the parameters observed in neutron β-decay (Close 1979).

4. BEYOND QUARKS AND LEPTONS

In most of what we have seen so far the leptons and quarks appear to form two independent families. However, there are clues that leptons and quarks are related. We have already commented that β-decays is essentially identical for quarks and for leptons. This includes the property that the two sets of leptons have electric charges that differ by one unit, as do the two sets of quarks. We might have thought *a priori* that the quark's electrical charges differed by some other amount.

In turn we have the remarkable fact that atoms are electrically neutral. The scale of charges on the quarks compares precisely with the leptons. Any difference between the lepton and proton charges is trifling (Close 1984):

$$Q_e + Q_p/Q < 10^{-19}.$$

Somehow the leptons and the quarks must 'know' of each other. The electrical charges not only have the same scale but are exactly equal. The proton has identical charge to the electron, rather than twice as much, because of the conspiracy between third-fractional charges and the threefold quark presence in the proton. This latter is a result of the colour forces saturating at three. If this is not to be dismissed as a mere coincidence it implies that the electromagnetic and colour forces are not independent.

During 1983 the discovery of the W and Z carriers of the weak forces confirmed that the electromagnetic and weak forces are related (A. Clark this symposium; J. Dowell, this symposium). Combined with the hint that electromagnetic and colour forces are also related, this in turn suggests the possibility that all of the three are united. The mathematical theories of these three forces have a common seed: they are all gauge theories. Grand unified theories combining all three forces have been constructed (J. Ellis, this symposium). One of the consequences is that the particles of matter – the leptons and the quarks – are in turn related. New forces are predicted which act over exceedingly short distances and which can transmute quarks into leptons. This can cause the proton to decay, a phenomenon that is being actively sought (T. Jones, this symposium).

To date, the GUTs have no satisfactory explanation for the triplication of the quark and lepton generations. Partly for this reason, some workers are pursuing the possibility that quarks and leptons are in turn composite particles.

We have qualitatively mentioned the problems with the lack of radiative transitions. These problems are perhaps sharpest when we study the electromagnetic interaction of the individual leptons.

The interactions of electrons and muons have been studied at momentum transfers approaching 100 GeV, corresponding to a resolution scale of 10^{-16} cm. No hints of structure have emerged, so the intrinsic size of any constituents must be smaller than this, five orders of magnitude smaller than the electron's Compton wavelength,

$$m_e^{-1} \approx 4 \times 10^{-11} \text{ cm}.$$

So if the electron is composite, it is unlike all previous examples.

For a typical non-relativistic system such as an atom, or a nucleus, the size R, mass M and binding energy E_B are approximately related by (Brodsky & Drell 1980):

$$R \approx (ME_B)^{-\frac{1}{2}} > M^{-1}.$$

Elementary quarks in a bag model give the nucleon a mass, M, related to the bag radius,

$$R \approx M^{-1}.$$

For the electron, by contrast,

$$R \ll M^{-1}.$$

Any composite model must explain why the electron and muon (and probably quarks) are so light on the scale of their sizes: $R^{-1} > 100$ GeV.

Perhaps surprisingly, precision experiments at low energy, measuring the magnetic moments of the electron and muon, place even more powerful constraints on composite models. The gyromagnetic ratios (g) of the electron and muon are very close to the Dirac value of 2, and there is no *a priori* reason for $g \approx 2$ for composite fermions.

The experimental value for $\frac{1}{2}(g-2)$ is determined most precisely for the electron and is

$$1\,159\,652\,209\,(31) \times 10^{-12}.$$

In QED these Dirac particles receive an anomalous magnetic moment from virtual quantum fluctuations. In a *tour de force* these have been computed up to order eight in perturbation theory (Kinoshita & Lindquist 1981). Theory and experiment agree to more than ten significant figures!

There is a possible discrepancy, however, for the electron:

$$\Delta a \equiv \tfrac{1}{2}(g-2)_{\text{th-exp}} = -270 \times 10^{-12},$$

if

$$\alpha^{-1} = 137.035\,963\,(15).$$

The error in the experiment is 31×10^{-12}; that on the theory is estimated to be 180×10^{-12}, of which two-thirds is a result of the experimental uncertainty in the value of the fine-structure constant, α^{-1}. This may be improved as the quantum Hall effect becomes better understood. At present, α^{-1} is best determined by the Josephson effect; compare with the quantum Hall magnitude (von Klitzing *et al.* 1980),

$$\alpha^{-1} = 137.035\,968\,(23).$$

Composite systems will, in general, have anomalous magnetic moments if their constituents are charged. Thus the proton and neutron, whose charges are in the ratio 1:0 have magnetic moments almost in the ratio $-\frac{3}{2}$. This ratio is well understood in terms of the quark content of the nucleons (Close 1979).

The absolute magnitude of the anomalous moment may be expected to be $O(mR)$ where m is the electron mass and R the intrinsic size of its constituents (Brodsky & Drell 1980). Given the limit (from high-momentum experiments) that $R < 10^{-16}$ cm, we anticipate that

$$\Delta a < 3 \times 10^{-6}.$$

As we have seen above, any non-zero value for Δa (after the QED contributions have been taken into account) is at least four orders of magnitude smaller than this. The limit on R is correspondingly reduced to $R_e \leqslant 2 \times 10^{-20}$ cm. For the muon, which is 200 times more massive than the electron, the limit is slightly better:

$$R_\mu < 10^{-20}\ \text{cm}.$$

The mass scale for the constituents is correspondingly

$$M > 2 \times 10^6\ \text{GeV}.$$

This estimate comes about if the lepton is a tightly bound composite of constituents whose masses $M \gg m_e$, and if we ignore any possibility of cancellations due to the symmetries in the underlying dynamics. Thus (Brodsky & Drell 1980),

$$\Delta a = O(m_e/M).$$

However, in a chirally invariant model, both left-handed and right-handed constituent fermions enter with equal weight in the state wavefunction of the composite lepton. Invariance under $M \rightarrow -M$ eliminates linear dependence on M (Brodsky & Drell 1980) and leads to

$$\Delta a = O(m_e^2/M^2).$$

In such models, $M > 870$ GeV (Eichten *et al.* 1983).

Lepton substructure is unlikely to be directly manifested at experiments below 10^3 GeV (1 TeV). If compositeness is revealed then there should also exist other combinations of the constituents. Thus we would expect to find coloured leptons, strong interactions among electrons, and strongly bound states of two or more electron (Bars 1982).

However, there could be effects of compositeness that will be indirectly manifested at lower energies. If the electron and positron are composite, they can interact by exchanging their constituents (analogous to covariant atomic forces). QED predicts specific angular distributions for the scattered leptons; constituent exchange could cause observable deviations from these distributions at energies as low as $\sqrt{s} = 100$ GeV in electron–positron collisions (Eichten *et al.* 1983).

Quarks and leptons are effectively structureless at present energies, and interactions among hadrons are analysed in terms of their constituent quarks. The question of lepton and quark substructure may be a challenge for future generations.

References

Bars, I. 1982 *Nucl. Phys.* B **208**, 77–121.
Brodsky, S. & Drell, S. 1980 *Phys. Rev.* D **22**, 2236–2243.
Burmester, J. 1977 *Physics. Lett.* **68**B, 301–304.
Chodos, A., de Grand, T., Jaffe, R., Thorn, C. & Weisskopf, V. 1974 *Phys. Rev.* D **9**, 3471–3495.
Close, F. E. 1979 *An introduction to quarks and partons.* London: Academic Press.
Close, F. E. 1983 *The cosmic onion.* London: Heinemann Educational.
Close, F. E. 1984 *Nature, Lond.* **309**, 513.
Cooper, A. & Sarkar, S. 1984 *Physics Lett.* **148**B, 347–352.
Cowsick, R. & McClelland, J. 1972 *Phys. Rev. Lett.* **29**, 669–670.
Dalitz, R. H. 1966 In *Proceedings of XIII international conference on high energy physics*, pp. 215–241. Berkeley: University of California Press.
Dicus, D., Kolb, E. & Teplitz, V. 1977 *Phys. Rev. Lett.* **39**, 168–171.
Eichten, E., Lane, K. & Peskin, M. 1983 *Phys. Rev. Lett.* **50**, 811–815.
Gell-Mann, M. & Ne'eman, Y. 1962 *The eightfold way.* New York: Benjamin.
Gershtein, X. & Zel'dovitch, Ya. B. 1966 *Soviet Phys. JETP Lett.* **4**, 120–121.
Kinnison, W. *et al.* 1982 *Phys. Rev.* D **25**, 2846–2869.
Kinoshita, T. & Lindquist, W. 1981 *Phys. Rev. Lett.* **47**, 1573–1576.
Lee, B. W. & Weinberg, S. 1977 *Phys. Rev. Lett.* **39**, 165–168.
Lubimov, V. A. 1980 *Physics Lett.* **94**B, 266–269.
Lubimov, V. A. 1983 In *Proceedings of HEP '83.* (Ed. J. Guy & C. Costain), pp. 386–389. Rutherford Appleton Laboratory.

Matteuzzi, C. *et al.* 1984 *Phys. Rev. Lett.* **52**, 1869–1872.
von Klitzing, K., Dorda, G. & Pepper, M. 1980 *Phys. Rev. Lett.* **45**, 494–498.
Vysotsky, M., Dolgov, A. D. & Zel'dovitch, Ya. B. 1977 *Soviet Phys. JETP Lett.* **26**, 188–189.
Wilkinson, D. H. 1981 In *The nature of matter* (ed. J. Mulvey), pp. 1–28. Oxford: Clarendon Press.

Discussion

J. H. MacGibbon (*University of Cambridge*). How recent is the upper bound on the τ-neutrino mass quoted in Dr Close's address?

F. E. Close. The upper bound quoted in the text was reported in February 1984. A limit of 70 MeV has recently been claimed by the ARGUS collaboration (Albrecht *et al.* 1985).

Reference

Albrecht, H. *et al.* 1985 *Physics Lett.* **163**, 404–407.

T. McMahon (*University of Birmingham*). Given the explanation of the proton rest mass in terms of the uncertainty principle and confinement of three quarks in a small volume, how is it possible to speculate about substructure in leptons, given their apparent point-like size and known low rest masses?

F. E. Close. In my opinion you have cited the major problem for builders of composite models. The fact that one does not know the dynamics should not deter one in studying the symmetry properties of models. Moreover, the 'point-like' size is only an experimental limit (the atomic nucleus appeared point-like initially).

Proc. R. Soc. Lond. A **404**, 167–188 (1986)
Printed in Great Britain

Jets: the materialization of quarks and gluons

BY R. MARSHALL
Rutherford Appleton Laboratory, Chilton, Didcot, Oxfordshire OX11 0QX, U.K.

Although quarks materialize in the observable world as jets of particles via a fragmentation process that is not perfectly understood, it is nevertheless possible to study quark scattering processes as if they were free particles. This paper, which is aimed at scientists outside the immediate field of particle physics, describes some of the properties of jets and how the jet observables can be related to quark parameters. The similarity of quarks and leptons is underlined.

This article is not meant for particle physicists who work daily with leptons, quarks, and related particles. But it is meant for scientists, and hence uses scientific language to describe what is known about jets, which are the materialization of quarks and gluons.

Every branch of science has a habit of developing its own particular vocabulary or jargon, a curious habit as it invariably hinders the flow of knowledge across the boundaries of the various fields. Consider also the following (oversimplified) logical sequence of the material structure of the chemical and physical world.

$$\begin{array}{c} \updownarrow \\ \text{polymers} \\ \updownarrow \\ \leftrightarrow \text{crystals} \leftrightarrow \text{molecules} \leftrightarrow \text{atoms} \leftrightarrow \text{nuclei} \leftrightarrow \text{nucleons} \leftrightarrow \text{leptons, quarks} \leftrightarrow \\ \updownarrow \\ \text{biomolecules} \\ \updownarrow \end{array}$$

Most items in this chain have a direct and obvious scientific relevance to adjacent members except at the micro end of the chain, where leptons and quarks represent the smallest structure known. If this partial isolation, which arises because no-one knows anything about lepto-quark substructure, were to be coupled with language obscurity, there would be a danger of the subject appearing distant and irrelevant.

The aim of this article is therefore to present the underlying simplicity of the subject of quark and gluon jets. Some specialized language is unavoidable, but it is kept to a minimum. Rather than using analogies which are frequently less clear than the original subject itself, direct comparisons are made which demonstrate that the scientific and mathematical techniques used to study the field are no different to those used in, say, X-ray crystallography, botany, or astronomy, even though the objects under scrutiny are many orders of magnitude smaller.

Introduction

The proton, being the nucleus of the hydrogen atom, as well as a constituent of all known atomic nuclei, itself contains several 'nuclei' or 'structural centres', called quarks. The quarks are manifestly there and constitute the structure of all known particles except the electron and its charged (μ and τ) and neutral (ν) leptonic relatives. The fact that all nuclei are of two types of building block, protons and neutrons, is a direct consequence of the fact that there are only two types of stable quark, the d and the u, with their unstable heavier relatives, the s, c, b and perhaps the t. And yet the quarks cannot be observed by any of the conventional measuring techniques which are and have been used to detect all other charged or neutral particles throughout scientific history. Whenever attempts are made to dislodge quarks from a proton (or neutron) or to 'ionize' the quark and antiquark which form the structure of mesons, then a 'stream' or 'jet' of particles (mesons, baryons, leptons, photons) emerges along the expected flight path of the quark or antiquark. The quark itself does not appear as an isolated identifiable particle, nor even as a minority member of the jet of particles; it remains confined by the potential which binds it into the proton. This is an experimental fact and hence is incorporated as an essential ingredient into the theory of the strong interaction – quantum chromodynamics (or QCD).

The phenomenon whereby quarks become jets of particles is somewhat misleadingly called 'fragmentation', implying that the quarks break up into fragments, which they do not.

The purpose of this paper is not so much to enquire why quarks are confined, nor even to consider the detailed mechanism of how quarks materialize into hadrons at the confinment interface (although we shall touch on these questions). Rather, we want to concentrate on the measurable properties of these jets and on how they relate to the basic physical properties of quarks, which is what we really want to study. It will emerge from these discussions, that although the fine details of how and why quarks materialize into jets are not fully understood, enough is known to be able to exploit jet production in many processes, so that the fundamental physical laws that govern the interactions of the quarks (and leptons) can be investigated.

There is also compelling evidence that the quarks contained in the so-called hadrons (the generic name for strongly interacting particles like the proton and the mesons which can be formed from quark bound states) are bound by forces that have some similarities to the well known electromagnetic force, but which also exhibit some important differences. The nature of these differences at present accelerator energies, and how these differences should disappear at (very) high energies, is discussed in the lectures by F. E. Close and J. R. Ellis. The relevance to the subject of jets is that the quantization of this strong force field leads to further particles analogous to the photon called gluons, and these gluons also appear in the observable world as jets. The interaction between two quarks (or quark and antiquark) can thus be described as the exchange of virtual gluons. Furthermore, quarks can emit gluons in a bremsstrahlung process similar to photon bremsstrahlung by electrons, and these real gluons also appear in the physical world as jets.

Table 1 shows the constituents of the various targets and projectiles used by particle physicists to study quark excitation, and the origin of the various jets that can be produced.

The recognition that the physical evolution of moving quarks is actually in the form of jets of particles was a gradual development during the 1970s. The initial difficulty in observing and hence measuring clearly defined jets was a direct consequence of the lower energies available for scattering and annihilation at accelerators in the mid-1970s compared with the present day. Although the energies for proton–proton collisions at the ISR accelerator at CERN were, in principle, high enough to provide a proton constituent with a large enough momentum to produce a jet, the accompanying collisions between protonic glue and the other quark spectators made the unambiguous detection of a clean quark jet rather problematic; the jets simply overlapped.

After the spectacular discovery in 1974 of the narrow J/ψ resonance in e^+e^- final states at Brookhaven and via e^+e^- annihilation at SPEAR (Stanford) together with

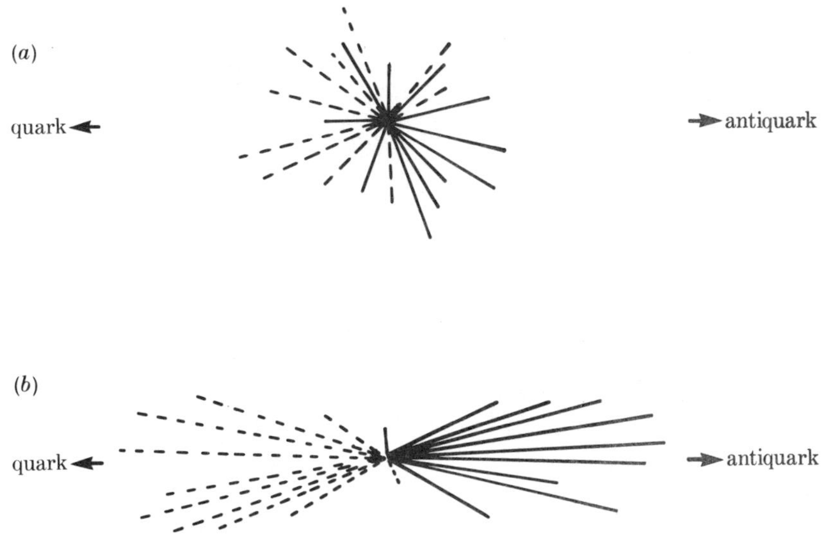

FIGURE 1. (*a*) A quark–antiquark, produced at low energy, fragments and produces an isotropic distribution of final-state particles. (*b*) A quark–antiquark, produced at high energy, fragments and produces two collimated, well separated jets.

TABLE 1. TARGETS, PROJECTILES AND JETS IN QUARK EXCITATION

target or projectile	content	jet types
baryons (p, n, etc.)	quarks plus gluons	quark jets diquark jets gluon jets
antibaryons ($\bar{p}$)	antiquarks plus gluons	antiquark jets antidiquark jets gluon jets
mesons photon	quark–antiquark plus gluons	quark–antiquark jets gluon jets

(*a*)

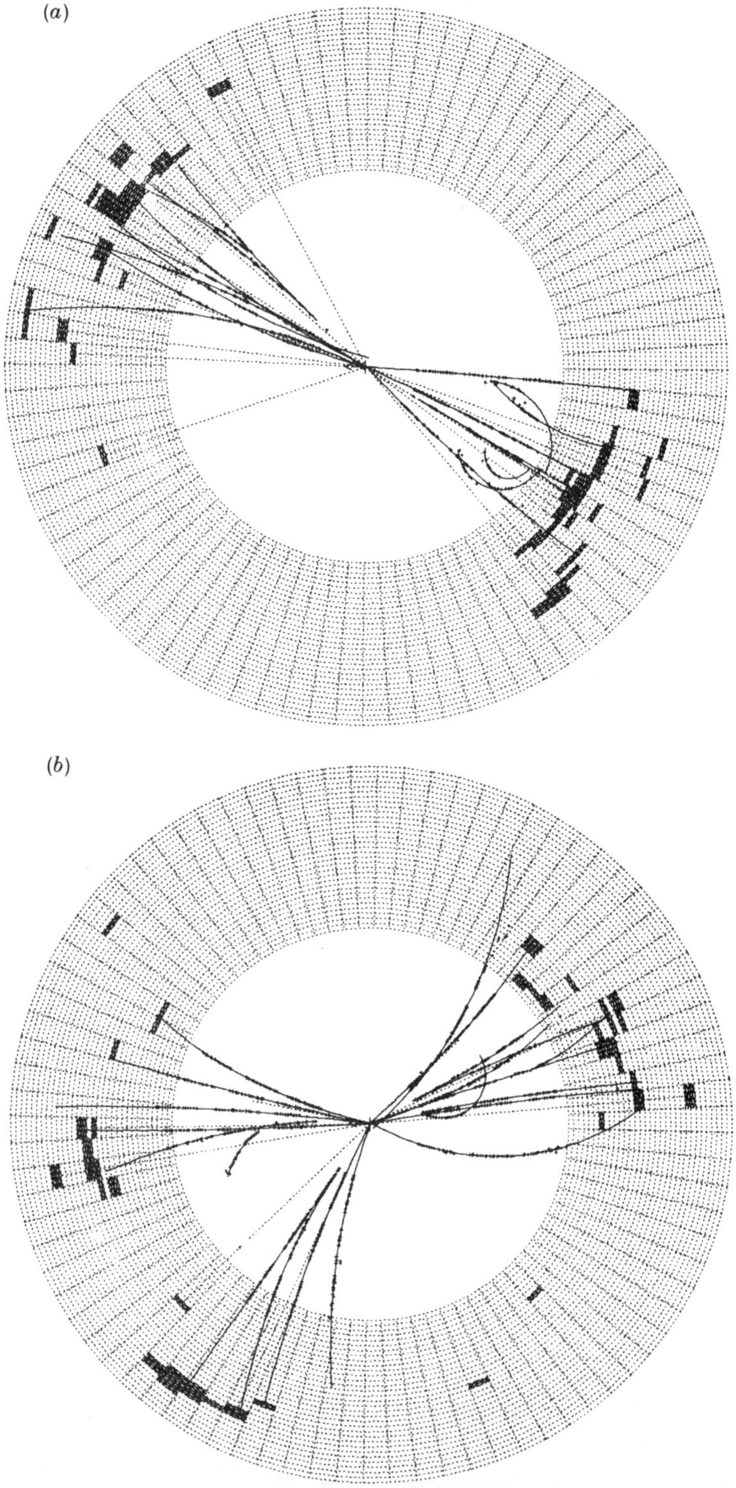

(*b*)

FIGURE 2. (*a*) A two-jet event (q–q̄) and (*b*) a three-jet event (q–q̄–g), observed by the JADE experiment at PETRA.

the subsequent detailed studies of this and other resonances, the experiments at SPEAR turned their attention to hadron production in e^+e^- annihilation in the continuum – away from the resonances.

There was clear evidence in these measurements for a non-spherical distribution of produced particles, and the interpretation was widely accepted that the quasi two-body process,

$$e^+e^- \rightarrow \text{quark (q)} + \text{antiquark } (\bar{q}) \rightarrow \text{hadrons},$$

was being observed.

Figure 1*a* shows the process graphically. The energies available at SPEAR were relatively low and the quark–antiquark tried to move apart with only a modest amount of momentum. As the quarks materialized into hadrons, the low momentum of the quarks meant that the materialization was rather isotropic; the jets were not particularly collimated and their existence was deduced from the overall deviation from isotropy rather than being observed directly.

However, it was already realized in 1975 that the energies of the new e^+e^- accelerators then being planned (PEP and PETRA) were so high, that if quarks really did materialize into jets as was strongly suspected, and the transverse momentum spread of the jets were essentially constant, then $e^+e^- \rightarrow$ quark–antiquark production at PEP and PETRA should be observed as two back-to-back, well collimated jets, aligned along the direction of the primary quarks, (see figure 1*b*). When PETRA started operation in 1978 and experimenters looked at their new data, precisely such configurations were observed (figure 2*a*). This process of e^+e^- annihilation into hadrons still remains one of the cleanest ways of producing quark jets for further study.

There is about a 20 % probability that either the quark or antiquark will radiate a hard gluon, leading to a three-jet configuration. Such events are also observed in the data and can easily be recognized as shown in figure 2*b*.

There are other ways, some of which can now produce jets of even higher energy and which can probe the quark interaction at smaller and smaller distances and which can produce clean samples of gluon jets.

Jet production mechanisms

To start the comparison of jet production mechanisms, figure 3*a* shows the annihilation of an electron and positron into two jets, which come from a quark and antiquark pair produced from the vacuum. Either the quark or antiquark can radiate one or more gluons, leading to three-jet and four-jet configurations.

A proton–proton collision can be drawn in a similar schematic fashion (figure 3*b*). The two quarks, which collide at high energy, undergo a large momentum transfer and try to leave the confines of the proton. They cannot do this, because of the confining bond, and so they radiate hadrons along the path they would have taken. In each proton, there are two 'spectator' quarks which receive a modest momentum transfer impulse and continue to travel in more or less their original direction. These diquarks also cannot exist in free space and they radiate a diquark

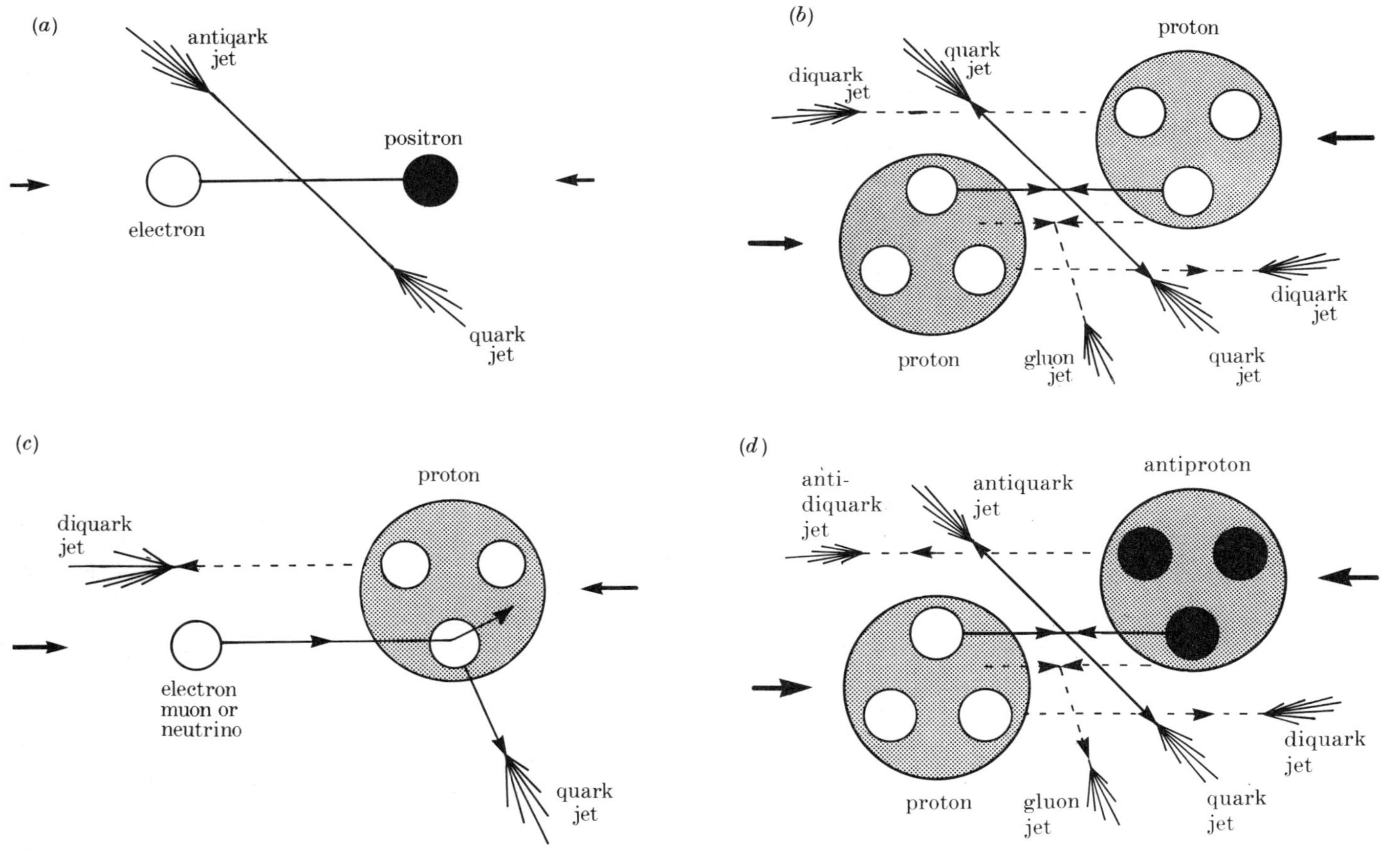

FIGURE 3. (a) Schematic diagram showing the annihilation of an electron and positron, producing a quark–antiquark which then fragments into jets. (b) Schematic diagram showing the collision between two protons. (c) Schematic diagram for lepton–quark scattering. (d) Schematic diagram for proton–antiproton scattering.

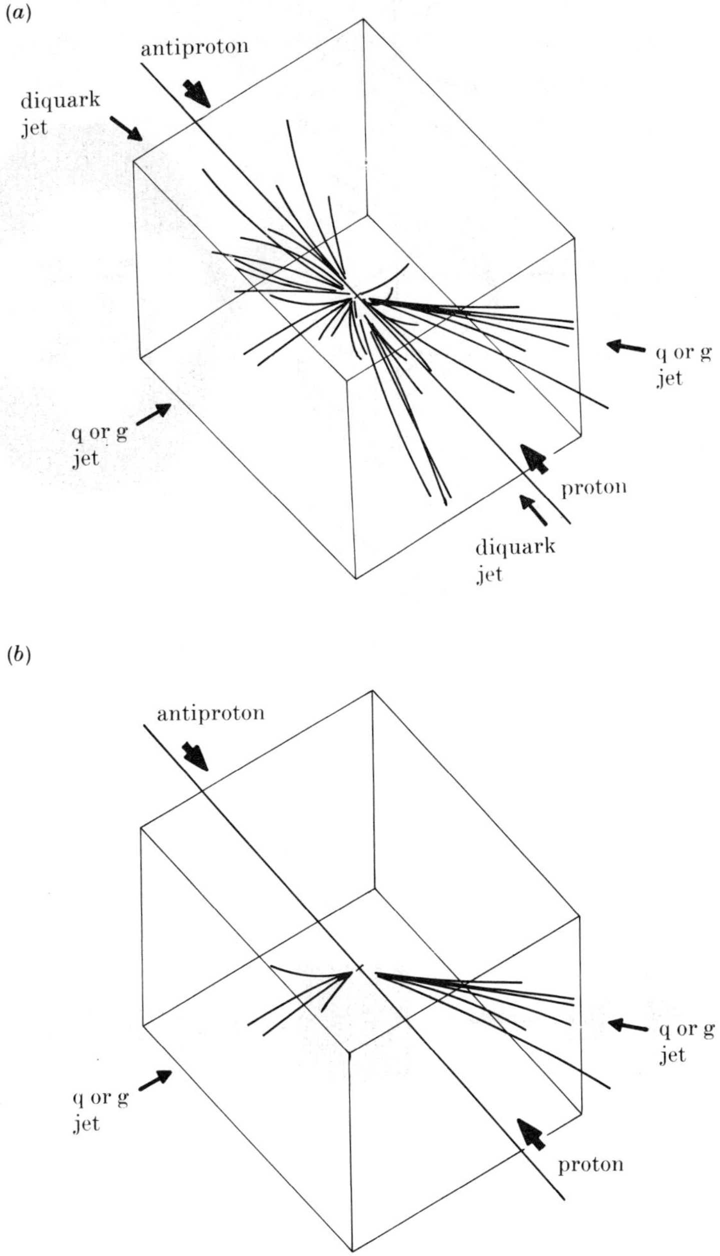

FIGURE 4. (*a*) The result of a $p\bar{p}$ collision measured by the UA1 experiment at the CERN collider. All particles produced in the collision are shown. (*b*) The same event with all particles removed which have a low transverse momentum component (less than 1 GeV c^{-1}) to the $p\bar{p}$ beam axis.

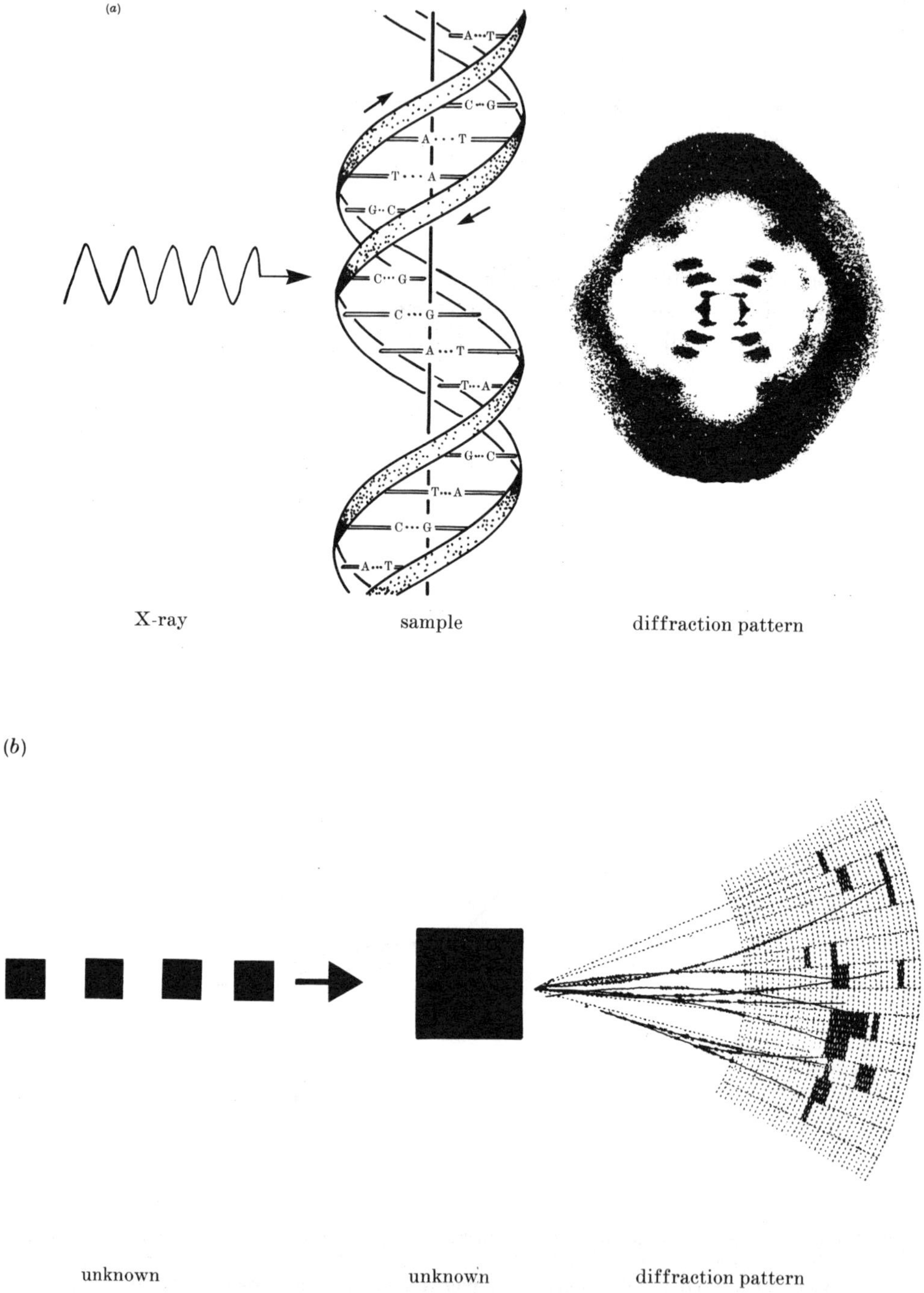

FIGURE 5. (*a*) Symbolic representation of an X-ray diffraction experiment. (*b*) Symbolic interpretation of jet production as a diffraction experiment.

jet along their flight path. A further process takes place. The protonic glue (shaded area in figure 3*b*) or gluon particles also interacts either with the gluons or quarks in the oncoming proton. These gluons are particularly interactive and in fact gluon–gluon collisions can dominate proton–proton interactions, leading to a proliferation of gluon jets. In such a collision, quarks, antiquarks, diquarks, or gluons (generically described as 'partons') can all produce their own particular jets.

Therefore, the result of a proton–proton collision can result in a large number of simultaneous jets of varying momenta which can overlap and merge, making observation of single, well identified, jets rather difficult. Nevertheless, the second generation of experiments at the ISR was able to embark on a thorough programme of jet physics during the early 1980s.

A substantial simplification is achieved if one of these colliding protons is replaced by an electron, e, (or muon, μ or neutrino, ν), as in figure 3*c*. There are now no gluon collisions to consider, because the gluon has been found to be invisible to lepton beams and effectively takes no part in the reaction. The struck quark produces a jet (the 'current' jet) and the residual diquark leads to the 'target' jet. At high energies, these two jets are well separated and from their directions with respect to the beam, a proton constituent jet (from either of the two u quarks or the d quark, which make up the proton) can be identified. There is still a 'background', however, because the electrons can emit a photon which then produces a quark–antiquark pair from the gluon sea.

The latest spectacular development since 1981, has been to collide a proton (p) and antiproton ($\bar{\text{p}}$) together (figure 3*d*). This is a kind of hybrid between e^+e^- annihilation (because the quarks and antiquarks annihilate) and pp collisions (because there are gluon and spectator end-products). The reason why this represents an advance lies in the ability to accelerate protons to far higher energies than electrons in circular accelerators. This is because electrons (due to their low mass) radiate vast amounts of energy as they are accelerated, in the form of synchrotron radiation, much to the delight of chemists, molecular biologists, and physicists, who can exploit this strong, polarized source, but rather to the consternation of particle physicts who would far rather see this energy in the form of particle motion. The highest energy of any primary accelerated electron beam in the world is about 30 GeV c^{-2}, whereas the highest energies for protons are more than 20 times higher. This is because the massive protons radiate a miniscule amount of synchrotron radiation.

Nevertheless, the result of a proton–antiproton collision can still contain substantial spectator diquark fragments together with the proceeds of gluon interactions. But now the collision energy is so high that the momentum transverse to the collision axis is extremely high for jets, owing to primary collisions. An event reconstructed from a p$\bar{\text{p}}$ collision is shown in figure 4*a*, where the various identified jets are indicated. The ability to study the physics of high momentum transfer collisions, where the jets emerge at wide angles, derives from the fact that particles in spectator jets have a low momentum component transverse to the collision axis because the spectators did not participate in the primary collision. By removing all particles with a transverse momentum (p_T) less than 1 GeV c^{-1}, the picture in figure 4*a* is transformed into that in figure 4*b*, where only the wide-angle jets remain. In this way the basic quark processes:

$$q\bar{q} \rightarrow \text{annihilation} \rightarrow \text{jets};$$
$$gg \rightarrow \text{scattering} \rightarrow \text{jets};$$
$$qg \rightarrow \text{fusion} \rightarrow \text{jets},$$

can be studied.

It is therefore apparent that any investigation of jets must concern itself with the origin of the jets, i.e. quark (or antiquark) jets, gluon jets, and diquark (or antidiquark) jets.

Moreover, there are six different kinds of quark: the charge $-\frac{1}{3}$ quarks, d, s and b; plus the charge $+\frac{2}{3}$ quarks, u, c and t, and the jets from these six different types of quarks are quite likely to be different. So, because of the multifarious origins of quarks it is clear that the subject of 'jet taxonomy' will play a part in unravelling these processes.

The basic physics problem is to deduce the kinematic and dynamic behaviour of the quark–gluon–diquark process via the physical observables, which are the measured particle vector momenta in the jet, and we shall investigate this problem shortly.

In a sense, a jet is a diffraction pattern, being a beam of scattered particles. The problem posed in figure 5*a* is well known to scientists where a well defined beam of X-rays of known wavelength scatter off an unknown sample, thus producing a measurable diffraction pattern. Knowing the parameters of the X-rays, and from the measured characteristics of the diffraction pattern, plus an intuitive leap, the structure of the sample can be deduced.

For the quark jet 'diffraction pattern', shown in figure 5*b*, neither the initial beam nor the scattering centre is known from the outset; both being essentially 'black boxes'. To deduce the properties of the moving quarks (equivalent to the X-ray beam) and the potential in which they move (the scattering sample) needs careful and detailed measurements of the jets themselves, together with information from many other sources such as the spectroscopic energy levels of the quarks in this potential (for example, baryon and meson masses). A few intuitive leaps are also needed.

Discrimination of quark jets

Some methods of classifying jets according to their origin via the measured characteristics of the jets themselves have their parallels in the taxonomy of plants.

Fisher's discriminant variable λ is well known in the literature of statistics. It is interesting to look briefly at the original application in Fisher (1936), to the classification of irises. Fisher obtained 50 samples each of three different species of iris, a total of 150 events. On each sample he measured the length and width (longitudinal and transverse dimensions) of the sepal (s_L and s_T) and length and width of the petal (p_L and p_T). The aim of the exercise was to determine four coefficients a_i ($i = 1, 2, 3, 4$), so that the single variable

$$\lambda = a_1 s_L + a_2 s_T + a_3 p_L + a_4 p_T$$

is the best discriminator for these irises, i.e. different species have the widest possible separation of λ values. Figure 6 shows the situation for irises.

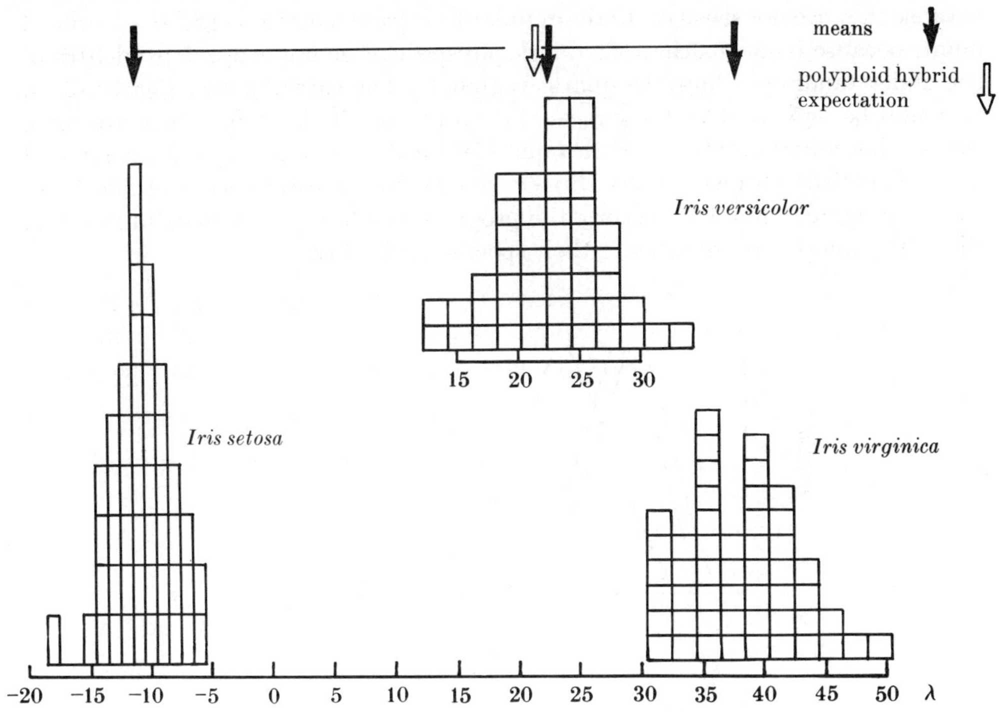

FIGURE 6. The variable λ plotted for three types of iris. One of the species is completely separated from the other two.

Rather similar methods have been used for jets (Marshall 1984). Here the width and length of jets, determined by the particles within a jet, have been used to classify jets according to their quark origin, allowing the physics of individual quark flavours to be studied.

An alternative method of quark jet identification uses the fact that the most recently dicovered quarks (c, b and t) are not only heavy: they decay with high probability via the weak interaction into lighter quarks. For example,

$$
\begin{array}{l} b \to cW \\ \qquad \;\; \hookrightarrow e\nu \\ \qquad \quad\;\; \mu\nu \\ \qquad \quad\;\; \tau\nu. \end{array}
$$

The decay electron (or μ or τ) has a large transverse momentum relative to the b direction, because the initial rest mass (energy) of the heavy b quark is converted into kinetic energy of the light decay products. It therefore doubly flags the heavy quark, i.e. the existence of e, μ or τ indicates a weak decay of a quark (plus background) and its high transverse momentum indicates a heavy quark.

For these studies, three variables were used to discriminate the quark species, chosen because they are sensitive to the quark mass. The variables are M_T, the overall (transverse) jet mass, $p_T(\mu)$, the transverse momentum of the μ and $p_T(\nu)$, the missing neutrino transverse momentum. Transverse components were chosen because they are not sensitive to the details of the fragmentation process – a crucial point, because it means that the quark physics can be studied independently of the understanding of how the quark fragments. The three lightest quarks, d, u, s, cannot be separated by these means; they are too similar. Thus there are three classes, dus, c, and b. (Studies of the t quark are still in their infancy and are ignored here.) The distributions of these three variables for the three classes of quark are shown in figure 7 and demonstrate the progress which has been made in isolating the heavy quarks, to investigate their specific properties.

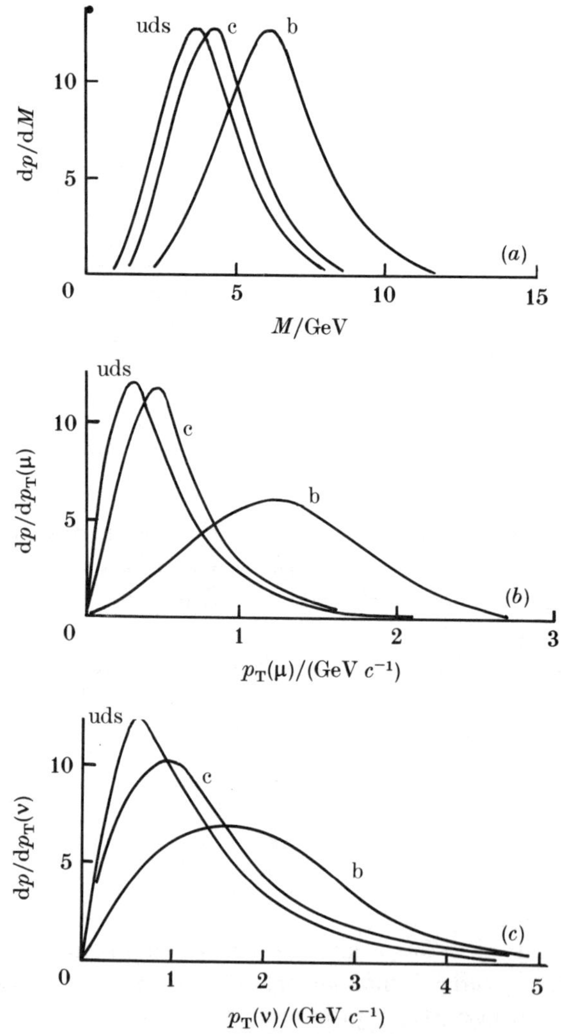

FIGURE 7. The variables (a) M_T, (b) $p_T(\mu)$ and (c) $p_T(\nu)$ plotted for three classes of quark. Quarks are not so easily separated as irises.

Fragmentation of light and heavy quarks

Although there is no definitive theory of the fragmentation process, several models have been proposed which at present provide a working phenomenological description of jet properties. The bulk of the particles in a jet are mesons, which are $q\bar{q}$ bound states. The models therefore present mechanisms for how these quark–antiquark pairs are produced from the vacuum.

It is possible to make generalized predictions without knowledge of the detailed mechanism, on the basis that some of the quarks are light (d, u, s) and some are heavy (c and b). Figure 8 sketches the early stages in the formation of a jet. The

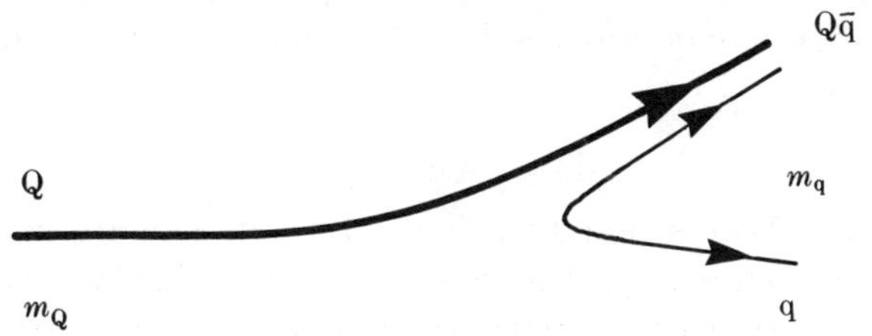

FIGURE 8. A schematic representation of the final stages of jet formation. The initial quark Q causes a $q\bar{q}$ pair production leading to a meson ($Q\bar{q}$). The other quark q continues the cascade.

primary quark Q of mass m_Q has been excited into motion by means of an external collision. As it tries in vain to move away from the confining force, the interaction causes pair production $q\bar{q}$, and one member of this pair joins with the original quark Q to produce a meson ($Q\bar{q}$), which carries a fraction z of the primary quark energy E_Q. The process can be thought of as a collision between the initial quark of mass

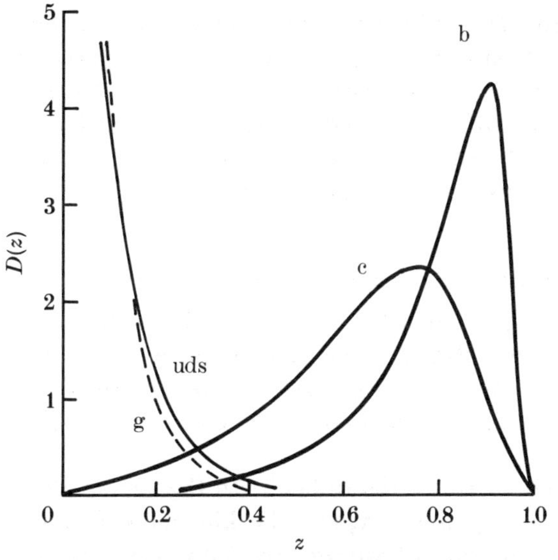

FIGURE 9. The fragmentation functions $D(z)$ for gluons and quarks.

m_Q and the quark–antiquark pair of mass m_q. Simple inertial arguments lead to the conclusion that the heavier the quark Q, the greater the fractional energy z carried by the meson containing the primary quark. In other words, heavy quarks should produce a harder spectrum of primary mesons than light quarks.

Extensive measurements have been made of the distribution functions $D_q(z)$ which describe the probability that the meson which contains the primary quark carries a fraction z of the primary quark energy E_Q. Figure 9 shows the measured distributions for light (d, u, s) quarks, c quarks ($m_Q \approx 1.5$ GeV c^{-2}) and b quarks ($m_Q \approx 4.5$ GeV c^{-2}). The results support the inertial arguments that the quarks behave as particles of reasonably well defined mass before the full process of fragmentation intervenes and converts the quark into a jet with a statistical distribution of kinematic quantities, like jet mass and kinetic energy.

Diquark jets

Diquark jets have been studied the least. In fact they are mainly regarded as an undesirable background in proton (or antiproton) collisions, and as we have seen in figure 4, efficient ways have been derived to eliminate them. However, it is quite possible that diquark jets will become increasingly important as a means of helping to understand the process by which quarks materialize and fragment into jets.

Quarks bind together either as a $q\bar{q}$ pair into a meson, or as a qqq triplet into a baryon. When quark physics (some time in the future) has become conventional quark chemistry with a perfectly understood theory for the forces which bind the particles, we will understand whether the q–$\bar{q}$ bond in mesons and the q–(qq) bond in baryons are the same, or different, and why. In chemistry, the unique coulomb force combined with the various structures of atoms and molecules gives rise to a wide range of apparently different forces or bonds, namely, ionic, covalent, metallic, hydrogen, Van der Waals, etc., which are all manifestations of the single electromagnetic force. It would therefore be no surprise if the unique QCD force plus the two different structures $q\bar{q}$ and qqq gave rise to different quark bonds, and hence different types of jets. For the moment, it is a matter for experiment to determine if these differences exist, although it will need rather delicate measurements to decide whether or not a quark jet from a (q–$\bar{q}$) situation is the same as a quark jet from a (q–qq) environment.

How to tell quarks from gluons

The different 'flavours' of quarks d, u, s, c, and b, produce measurably different jets, allowing the different flavours to be discriminated. Even without the stimulus of model predictions, which indicate that quarks and gluons (g) should produce different kinds of jet, natural curiosity would lead one to look at the data to see if this were true.

To identify a gluon jet is not too difficult. In the simple gluon bremsstrahlung process, the simplest way is to exploit the kinematic fact that in the process $e^+e^- \rightarrow q\bar{q}g$, the gluon jet has a lower energy than either the q or the $\bar{q}$ jet about 80% of the time. Even after experimental smearing in a typical detector, this is still true in about 60–70% of examples. So, by selecting the lowest energy jet in each event, an enriched sample of gluon jets can be obtained. Another source of gluon jets is in $p\bar{p}$ scattering at the CERN $p\bar{p}$ collider. In these very high energy collisions, most of the jets are from gluons and it is relatively easy to collect a pure sample of gluon jets for further study.

Both e^+e^- and $p\bar{p}$ experiments lead to the same conclusion: gluon jets are noticeably different from quark jets of the same energy. The e^+e^- investigations concentrated on the average transverse momentum of all particles in a jet. For a given jet energy, this transverse momentum average is always larger for the enriched gluon jet sample than for quark jets. This means that quark jets are physically broader than a quark jet of the same energy.

The $p\bar{p}$ results concentrate on the gluon fragmentation function $D_g(z)$, which is essentially the longitudinal momentum dependence of the primary particle in the jet. By momentum and energy conservation, this is the complement of the transverse momentum studies mentioned above. The results of the UA1 measurements are shown in figure 9 alongside the various quark functions and lead to the same result as the e^+e^- measurements. Gluon jets are slightly broader than quark jets and contain particles which on average have a lower momentum.

Jets as parton estimators

There is considerable experimental evidence to support the contention that the summed relativistic four-momentum of all particles in a jet is a useful statistical estimator for the primary parton four-momentum. Under some circumstances, this estimator for the vector properties of the parton may be less accurate than if the jet had appeared in the detector as a single charged particle. However, sometimes it can be more accurate and the fact that the quark or gluon has fragmented actually makes it easier to measure its kinematic properties. This is simply a function of the jet energy and detector resolution and a simple example demonstrates the situation clearly. A typical tracking detector for charged particles has a momentum resolution σ_p, which is proportional to p^2, where p is the momentum itself:

$$\sigma_p = kp^2.$$

The resolution for a hypothetical parton of momentum p_0 is thus kp_0^2. Suppose the parton actually produces a jet of N particles each with momentum p_0/N. (Actually there would be a statistical distribution of momenta with average p_0/N, but this does not change the results of the argument.) The resolution for the individual components is then kp_0^2/N^2 and if these are then combined together to estimate the parton momentum, the error on the measurement from the detector is

$$\sigma_{p(\text{jet})} = kp_0^2/N^{\frac{3}{2}},$$

which is considerably smaller than the resolution for a single parton of equivalent momentum. It is ironic that if quarks appeared as singly charged particles without fragmenting, the fact that they have fractional charge means that they would bend even less in the magnetic fields of the detectors with a worse resolution than that for the single particle derived above. In this context, because the quarks fragment, their momentum can, in principle, be measured more accurately than if they did not fragment.

To see how quark and gluon kinematics work in practice, it is instructive to compare lepton and quark production, particularly in e^+e^- annihilation.

Consider the processes for lepton production where the final and initial states are not identical:

$$e^+e^- \rightarrow \mu^+\mu^-,$$
$$e^+e^- \rightarrow \tau^+\tau^-,$$

and the corresponding reaction for quarks:

$$e^+e^- \rightarrow \text{quark–antiquark} \rightarrow \text{jet} + \text{jet}.$$

In all these examples, the reaction proceeds, in theory, via an intermediate virtual γ with a well defined angular momentum ($J^P = 1^-$), and this leads to a specific prediction for the angular distribution for the μ^+ (say) with respect to the e^+ direction, namely

$$d\sigma/d\Omega \sim (1+\cos^2\theta).$$

To be precise, there is an additional probability that the reaction proceeds via the recently discovered weak boson Z^0 ($J^P = 1^-$ and 1^+) and the subsequent interference between the 1^+ and the 1^- leads to a small $\cos\theta$ term which can be extracted from the data (see figure 10 for $e^+e^- \rightarrow \mu^+\mu^-$).

A similar angular distribution can be measured for jets arising from quark–antiquark production. At this point, let us not enquire which jet is the quark and which is the antiquark, but simply plot the folded angular distribution compared with the unfolded distributions in figure 10. (Now the $\cos\theta$ term disappears in the folding.) The result is figure 11, where a comparison with the expected $1+\cos^2\theta$ shape is made. Good agreement can be seen. Figure 11 is consistent with a far-reaching interpretation of jet production as follows:

(i) the jets accurately estimate the primary quark direction;
(ii) the quark dynamics can be deduced from the jet properties;
(iii) the primary quarks have spin $= \frac{1}{2}\hbar$ – like the leptons;
(iv) the process $e^+e^- \rightarrow q\bar{q}$ proceeds via a virtual photon (γ), i.e. quarks can be pair-produced by photons, just like $\gamma \rightarrow e^+e^-$.

Simple experiments like these help to build up a picture where, apart from their strong interaction which manifests itself in fragmentation (among other things), quarks have a striking similarity to leptons.

By analogy with studies of the dynamics of low-energy nuclear particles, it would be an important step forward to be able to measure in any quark scattering process, the polarization or spin orientation of the quarks. It is not yet known how to do this. The question of whether the polarization of the quark may be deduced from the construction of pseudoscalar triple products from the jet constituents is still open. Further studies will be needed before we have any practical method for the measurement of quark spin polarization.

10

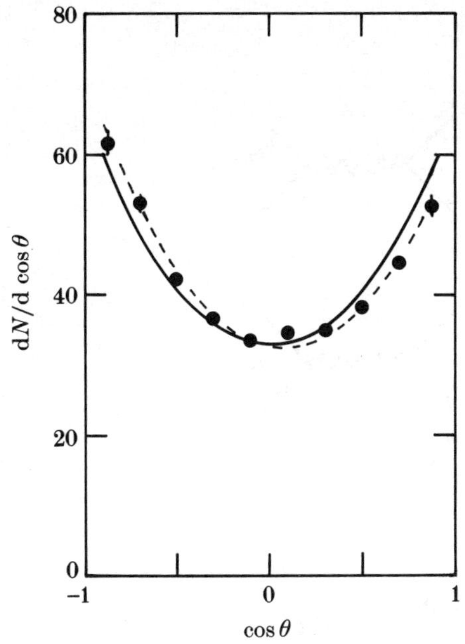

11

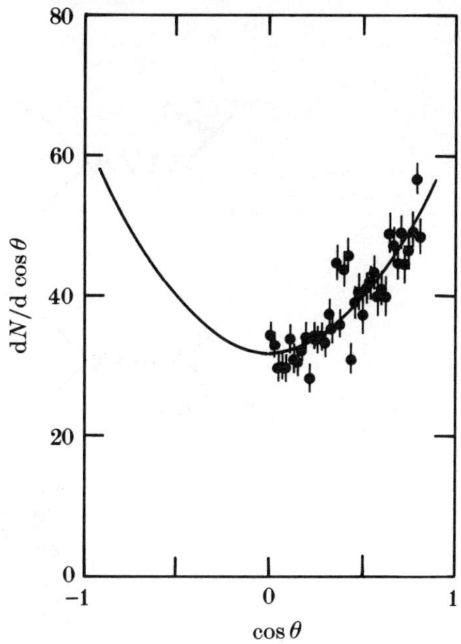

FIGURE 10. The angular distributions $d\sigma/d\Omega$ for the processes $e^+e^- \to \mu^+\mu^-$ (MAC). The dominant $(1+\cos^2\theta)$ term is apparent (full line), together with a noticeable $\cos\theta$ term from the electroweak interference.

FIGURE 11. The folded angular distribution $d\sigma/d\Omega$ for the process $e^+e^- \to q\bar{q}$ (JADE). The measured distribution agrees well with the expectation $(1+\cos^2\theta)$, shown by a full line.

Some rather nice γ interference experiments have been done recently by Ford *et al.* (1983) and Bartel *et al.* (1984), which also demonstrate clearly how quarks behave like charged leptons, underneath their fragmentation disguise. Here, the lepton reaction is similar to the pair-production process considered above, but with the addition of an extra real photon which is detected by the apparatus,

$$e^+e^- \to \mu^+\mu^-\gamma,$$
$$e^+e^- \to \tau^+\tau^-\gamma.$$

The measured photon can be radiated either by the initial e^+e^- or by the final state $\mu^+\mu^-$ or $\tau^+\tau^-$. The initial- and final-state particles have a definite phase relation and so the two sources of the radiated photons can and do interfere. Figure 12*a* shows the two graphs for these reactions, which interfere with each other and produce an asymmetric term in the angular distribution. Figure 12*b* represents the situation as a double-slit experiment where the slits are crossed at an angle θ instead of being parallel. The photon can then be emitted either from the lepton line or from the quark line (the two slits) and then interfere at the detector. If the slits were parallel, then interference fringes would be seen with a modulation in the direction of the slit separation; but if the slits are crossed, then it should not be surprising that the modulation is a function of the slit separation θ. A full calculation with quantum electrodynamics (QED) produces the result that the modulation goes through one cycle as θ goes from 0° to 180°, with a minimum of

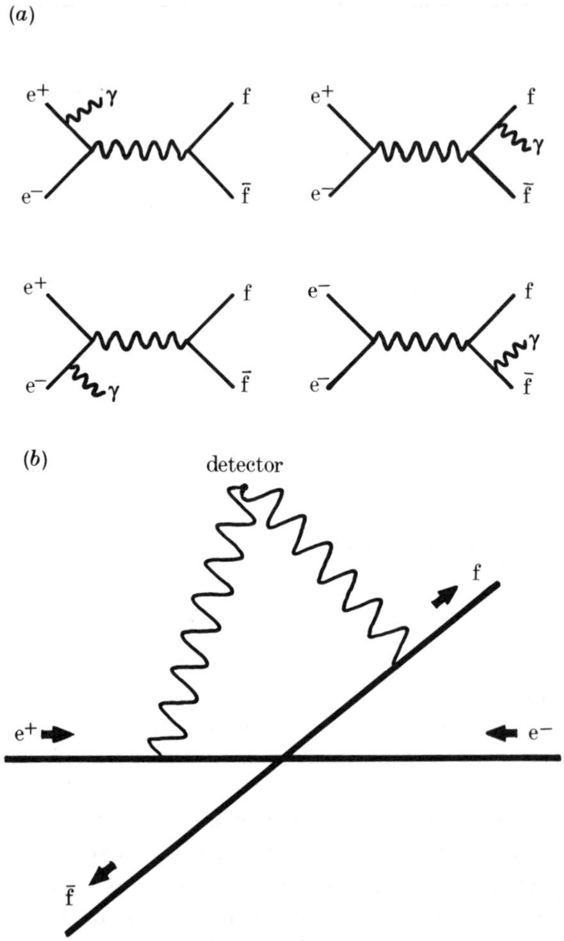

FIGURE 12. The representation of the process $e^+e^- \rightarrow f\bar{f}\gamma$ ($f = \mu$ or τ) as (a) Interfering diagrams. (b) A double-slit interference experiment.

0° and a maximum at 180°. This is precisely what is seen in the measurements (see figure 13). The fall-off to zero at 0° and 180° is an instrumental effect; it is not possible to mount a detector in the path of the e^+e^- beams.

The situation is slightly more tricky for jets. The reaction is now

$$e^+e^- \rightarrow \text{jet jet}\,\gamma,$$

and the first difficulty is to find the radiated photons in the jet debris, which also contains about 10 γ per event from π^0 decays. The π^0 photons are usually softer, but there is some overlap and confusion, although this is not serious. In addition, the radiated γ tend to be more isolated from the jet, whereas γ from π^0 are contained well within the jet. The second difficulty is in identifying the quark charge via the jet charge. Extensive simulation studies have shown that the summed jet charge is a statistical estimator for the quark charge, although it is not perfect (the jet charge has the same sign as the quark charge only about 70 % of the time. With these difficulties wholly or partly overcome, a result was obtained for quarks, as shown in figure 14. There is evidence for interference, even though higher statistical accuracy from further studies is desirable.

13
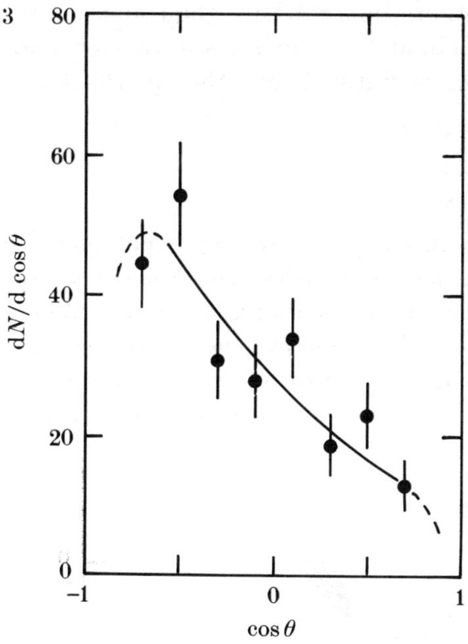

14
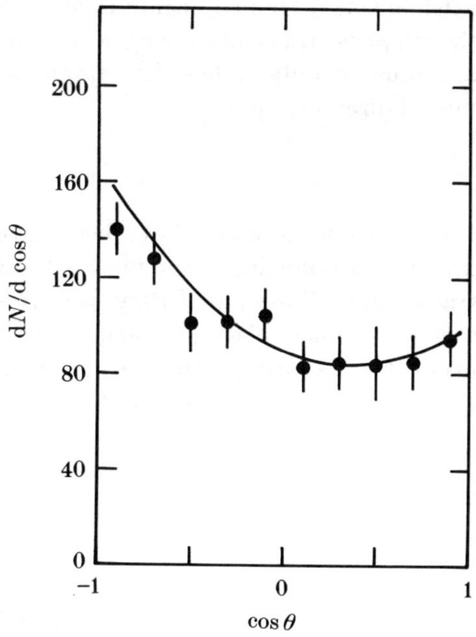

FIGURE 13. The angular distributions for the process $e^+e^- \rightarrow e^+e^- \rightarrow \tau^+\tau^-\gamma$ (MAC), showing the existence of an interference pattern.

FIGURE 14. The angular distribution for the process $e^+e^- \rightarrow q\bar{q}\gamma$ (JADE), showing a possible interference pattern.

THE SIZE OF THE JET SOURCE

In the 1950s, a series of stellar interference experiments were done at Manchester by Hanbury Brown and co-workers, indicating a new way to determine the size of stellar objects. The results were initially controversial because the interference took place between intensities rather than amplitudes (contrary to previous experience and intuition) until it was appreciated that the interference effect was a natural consequence of quantum mechanics. The qualitative characteristics of the 'Hanbury Brown–Twiss' effect are such that if a photon from a distant extended source is observed in a detector, there is an enhanced probability that a similar photon (same wavelength) will arrive at the detector at the same time. If two such detectors have a separation d and measure the time interval δt between photons with energy (wavelength) separation δE, then it is found that the probability falls off with d, δt and δE and the rate of fall-off is an indication of the size of the radiating source.

By reducing the astronomical scale by about 25 orders of magnitude, similar interference effects have been sought and observed in the jet constituents, by using spinless bosons (pions) instead of photons. The results of the TPC experiment (1985) at PEP on 'Bose–Einstein' correlations indicate that the particles, and hence the jets, are emitted from a region with a size of *ca.* 1 fm. This is comparable with the size of the proton, which is effectively the confinement radius. It therefore provides some information on the quark → jet transition region. If the quark phase exists inside a volume of radius *ca.* 1 fm (the proton size), then the transition to the jet phase takes place with a skin also of about 1 fm thickness so that beyond a radius of only a few fm, fragmentation is finished and the quarks have materialized into jets.

Summary

Only a few examples have been given in this paper of the way that quarks, despite materializing in the physical world as jets of particles, nevertheless can be traced and followed as if they were freely moving particles like electrons. When a collision takes place producing a quark with high energy, the quark behaves practically as a free particle, interacting through its electric and weak charges and through the strong interaction. It behaves freely because at small quark separations

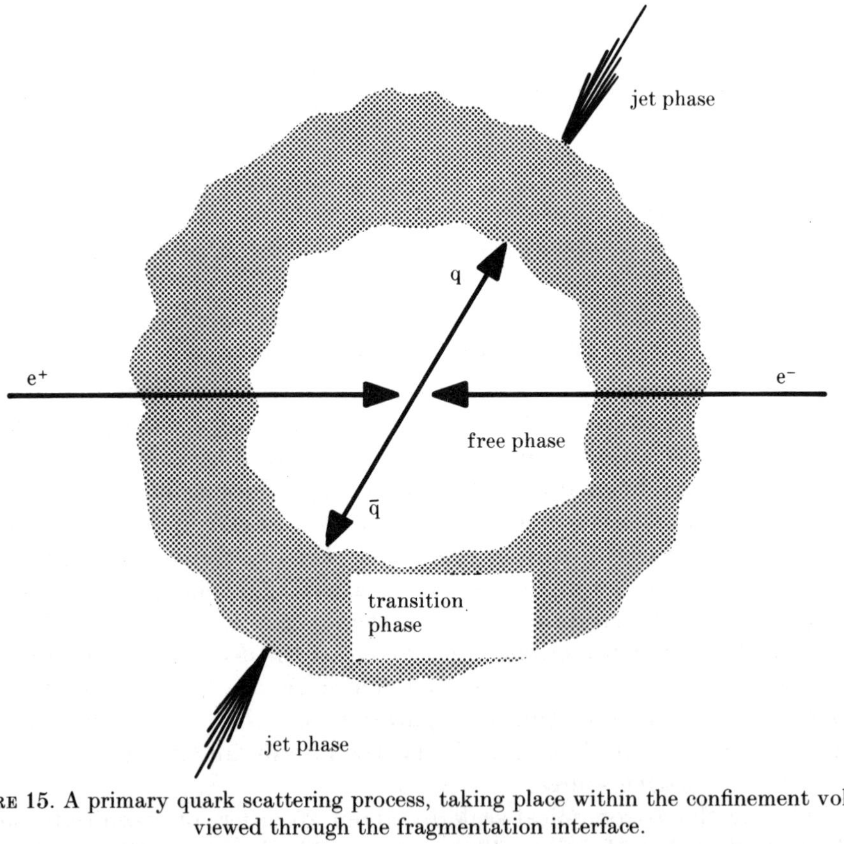

FIGURE 15. A primary quark scattering process, taking place within the confinement volume, viewed through the fragmentation interface.

and at high energies, both the coupling strength α_s and the potential are small. At a distance of about 1 fm, the kinetic energy begins to be used up in overcoming the ever-increasing potential energy caused by the quark trying to move out of the potential well. The jet is then produced effectively as a stream of radiation as the quark is slowed down by the retarding potential. Before the quark has moved much more than another 1 fm, the process is over and the jet is launched into the outside world. But the jet provides statistical estimators for what was happening inside the volume of radius 1 fm, and this allows us to observe the process that took place inside this sphere, albeit with reduced accuracy in some examples. Figure 15 presents such a scattering process, viewed through the fragmentation interface.

Table 2 summarizes the quark properties that we would like to observe, together with the status of how well these can be deduced at the present time.

TABLE 2. QUARK PROPERTIES

quark or gluon property	jet property	status
mass	jet transverse mass	good, but some overlap between various classes
quark flavour	jet transverse momentum	
quark or gluon identification	jet longitudinal momentum	
vector momentum	jet constituent summed momentum	accurate
energy	jet energy	affected by jet mass fluctuations and particle identification problems
spin	jet axis angular distributions	accurate, gives spin $\frac{1}{2}$ quarks and spin 1 gluons
polarization state	jet constituent triple products	nil

Because jets have only been studied for a relatively short period of time, normal experience would suggest that further improvements are to be expected, even in areas that have been comparatively unsuccessful so far. Nevertheless, enough has been established already about quark jets to permit considerable progress to have been made in the study of particle interactions of quarks. Even though the whole process of quarks $\rightarrow$ jets is imperfectly understood, this does not prevent quarks, as particles, being used to probe the structure of matter at distances less than 1/200 of the size of the proton. Thus, the sophisticated detecting apparatus which has been used in recent years to study jet production is nothing more than a quark version of the electron microscope, and these quark microscopes now have the ability, by exploiting jet production, to peer into the fine structure of the proton, the elementary particle of yesterday.

References

Aihara, H. *et al.* 1985 *LBL Berkeley preprint LBL*-18478 (31 pages).
Bartel, W. *et al.* 1984 *Zeit. f. Phys.* C **24**, 223–229.
Fisher, R. A. 1936 *Ann. Eugenics* **7**, 179–188.
Ford, W. T. *et al.* 1983 *Phys. Rev. Lett.* **51**, 257–260.
Hanbury Brown, R., Jennison, R. C. & Das Gupta, M. K. 1952 *Nature, Lond.* **170**, 1061–1082.
Hanbury Brown, R. & Twiss, R. Q. 1954 *Phil. Mag.* **45**, 663–682.
Marshall, R. 1984 *Zeit. f. Phys.* C **26**, 291–299.

Discussion

E. Gabathuler (*Liverpool University*). From the result Dr Marshall showed on the final state photon interference involving quarks, is it consistent with fractional $\frac{2}{3}$ charge?

R. Marshall. The MAC experiment actually measured the quantity $3\Sigma\langle E_Q\rangle\langle E_Q\rangle$ to be 1.97 ± 0.61. For fractionally charged quarks, this number is expected to be 2.11, whereas for integrally charged quarks, it would be 3.67.

Proc. R. Soc. Lond. A **404**, 189–211 (1986)
Printed in Great Britain

The physics of $W^{\pm}$ and Z^0 particles

By A. G. Clark
CERN, CH 1211 *Geneva* 23, *Switzerland*

The standard model is a theoretical framework describing the behaviour of elementary quarks and leptons as a result of strong and electroweak interactions. Our current understanding of the production and decay properties of the $W^{\pm}$ and Z^0 particles, the exchange bosons of the weak interaction, will be described and the striking agreement of these properties with predictions of the standard model will be emphasized.

1. Introduction

For many years, electromagnetic interaction processes have been successfully described by a gauge theory, quantum electrodynamics or QED, in which elementary fermions interact by the emission or absorption of the electromagnetic 'intermediate vector boson' (IVB), which is the quantum representation of the electromagnetic field. Here the IVB is the photon, γ.

During the past two decades, there have been three important developments.

(i) Our concept of the 'elementary fermion' has evolved. The known physical particles (for example, the proton, neutron, and pion) are now considered to be bound states of elementary quark constituents. Strong, weak, and electromagnetic interactions involving quarks can be described by the emission or absorption of the appropriate IVB. Leptons are thought to be elementary, and weak or electromagnetic processes involving leptons can be similarly described. No experimental evidence exists for strong interaction processes involving leptons.

(ii) The gauge theory of QED has been extended by Glashow (1962), Salam (1968), and Weinberg (1967) (the GSW model), and others, to include weak-interaction processes by the transfer of a triplet of 'weak' IVBs (the W^+, W^-, and Z^0 particles).

(iii) A similar gauge theory, QCD, has been developed to describe strong interaction processes by the transfer of an octet of 'strong' IVBs or gluons, which transfer 'strong' or 'colour' charge.

Original contributions to the QCD theory (see, for example, the review of Politzer (1974)) were soon followed by confrontations of the theory with experiment. Some aspects are reviewed by Darriulat (1984) and Di Lella (1985).

To understand weak-interaction processes from charged currents, the $W^{\pm}$ particles were expected long before the development of the GWS model. However, it was this model that required the existence of weak neutral currents. Experimental evidence soon followed with the observation of weak neutral currents (Hasert *et al.* 1973) in the scattering of neutrinos (and antineutrinos) from electrons or nucleons. It was only after the development of this model that quantitative predictions of the IVB properties were available.

Since the first direct observation of the $W^{\pm}$ and Z^0 particles at the CERN $\bar{p}p$

collider (Arnison *et al.* 1983*a*, *b*; Banner *et al.* 1983; Bagnaia *et al.* 1983), the UA1 and UA2 experiments have identified a total of about 300 $W^{\pm}$ events and 45 Z^0 events. This enables a quantitative examination of the validity of gauge theories in describing the production and decay of $W^{\pm}$ and Z^0.

This talk attempts a comparison of $W^{\pm}$ and Z^0 data with predictions of both q.c.d. and the G.S.W. model. In doing so, we largely ignore an extensive body of data in outstanding agreement with, and crucial to the understanding of, the strong and electroweak gauge theories. Several reviews exist (for example, Pullia (1984), Naroska (1983), and Langacker (1984)), in addition to the talks of R. Marshall and D. H. Saxon (this symposium). Section 2 qualitatively describe some aspects of the GSW model and QCD theory. Section 3 describes the data samples for $W^{\pm}$ and Z^0 presently available at the CERN collider, at $p\bar{p}$ centre of mass energies $\sqrt{s} = 546$ GeV and $\sqrt{s} = 630$ GeV. Sections 4, 5 compare aspects of the data with model predictions, and in §6 we summarize some aspects of the GSW model that have not yet been tested.

2. Comments on the GSW and QCD models

From our current knowledge of weak interactions, we can group the known fermions into three identical families containing left-handed (LH) doublets and right-handed (RH) singlets. We can assign a charge quantum mumber to each element of the group, and it is convenient to define a 'weak-isospin' quantum number whose third component, I_3, is shown in table 1.

Table 1. Quantum numbers for different fermion families

				I_3	Q	$g^{c}_{L,R}$	$g^{N}_{L,R}$
$\begin{pmatrix}\nu_l \\ l^-\end{pmatrix}_L$	$\begin{pmatrix}\nu_e \\ l\end{pmatrix}_L$	$\begin{pmatrix}\nu_\mu \\ \mu\end{pmatrix}_L$	$\begin{pmatrix}\nu_\tau \\ \tau\end{pmatrix}_L$	$\frac{1}{2}$	0	1	$\frac{1}{2}$
				$-\frac{1}{2}$	-1	1	$-\frac{1}{2}+x$
$\begin{pmatrix}q_1 \\ q_2\end{pmatrix}_L$	$\begin{pmatrix}u \\ d'\end{pmatrix}_L$	$\begin{pmatrix}c \\ s'\end{pmatrix}_L$	$\begin{pmatrix}t \\ b'\end{pmatrix}_L$	$\frac{1}{2}$	$\frac{2}{3}$	1	$\frac{1}{2}-\frac{2}{3}x$
				$-\frac{1}{2}$	$-\frac{1}{3}$	1	$-\frac{1}{2}+\frac{4}{3}x$
l^-_R	l_R	μ_R	τ_R	0	-1	0	x
q_{1R}	u_R	c_R	t_R	0	$\frac{2}{3}$	0	$-\frac{2}{3}x$
q_{2R}	d'_R	s'_R	b'_R	0	$-\frac{1}{3}$	0	$\frac{1}{3}x$

	g_A	g_V
ν	$\frac{1}{2}$	$\frac{1}{2}$
e, μ, τ	$-\frac{1}{2}$	$-\frac{1}{2}+2x$
u, c, t	$\frac{1}{2}$	$\frac{1}{2}-\frac{4}{3}x$
d′, s′, b′	$-\frac{1}{2}$	$-\frac{1}{2}+\frac{2}{3}x$

$x = \sin^2\theta_W$; $g^N_L = I_3 - Qx$; $g^N_R = -Qx$; $g_V = g_L + g_R$; $g_A = g_L - g_R$.

(i) Three multiplets are shown. Only the t quark has not been identified with certainty, although possible evidence for its existence at the collider has been presented (Dowell 1985). Within the electroweak model, no requirement is placed on the number of families.

(ii) It is a well justified assumption that charged currents act with identical coupling strength, g, on any lepton multiplet, and that the interaction involves only pairs of a given multiplet (lepton conservation). Furthermore, it is well verified from β-decay and neutrino interactions that the charged currents act only on left-handed doublets.

(iii) Weak neutral currents also act on right-handed singlets, and it is at this point that the GSW gauge theory makes specific predictions.

(iv) In the quark sector, the situation is identical, except that strangeness-changing baryon decays are observed experimentally. This is because the quarks (d′, s′, b′) of table 1 are mixtures of the 'physically observed' quarks (d, s, b). The quark mixing is described by the Kobayashi–Maskawa (1973) matrix. The current status is reviewed by Chau (1983) and Stone (1983), and is not considered further in this talk.

What the GSW model now assigns are charged and neutral coupling strengths to the l.h. and r.h. fermions in terms of the experimentally determined number $\sin^2\theta_W$. If Q is the fermion charge, and $x = \sin^2\theta_W$, the couplings are listed in units of the electromagnetic coupling $e\,(=1)$ in table 1.

These couplings can be derived as an extension of the QED gauge theory (the GSW model), and specific relations are made between e (or the electromagnetic coupling $\alpha_{em} \equiv \alpha$, the Fermi constant G_F, the weak mixing angle $\sin^2\theta_W$, and the mass, m_W, of the W$^{\pm}$ boson:

$$m_W = \frac{\pi\alpha}{\sqrt{2G_F}}\frac{1}{\sin\theta_W} = \frac{(37.2804 \pm 0.0003)}{\sin\theta_W}\ (\mathrm{GeV}\ c^{-2}). \qquad (1)$$

The effective relative strengths of the neutral and charged currents is defined in this talk by a fourth parameter (Ross & Veltman 1975; Hung & Sakurai 1978)

$$\rho = \frac{m_W^2}{m_Z^2\cos^2\theta_W}. \qquad (2)$$

Making a renormalizable gauge theory involving a triplet of exchange bosons has two consequences.

(i) The gauge theory is non-Abelian, with the consequence that the bosons self-couple (see, for example, figure 1).

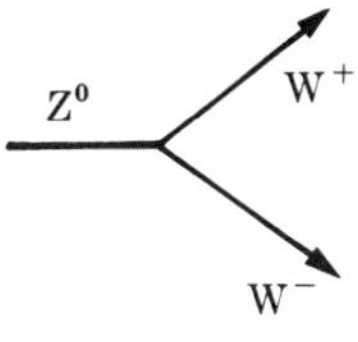

FIGURE 1. Example of self-coupling of the Z^0, W$^+$, and W$^-$ particles.

(ii) To maintain invariance of the electroweak theory in local gauge transformations, with massive m_W and m_Z, at least one additional scalar H (Higgs particle), must be introduced to cancel divergences introduced by the boson masses. In the simplest model of GSW, there is one Higgs scalar that is both self-coupled, and coupled to the W$^{\pm}$, and Z^0, and all fermions proportionally to their masses. Here (neglecting small corrections from higher order processes), $\rho = 1$. There is no experimental evidence for the existence of the Higgs boson.

These formulae must be modified for radiative corrections, either electromagnetic (for example, figure 2*a*), or weak (for example, figure 2*b*). Electromagnetic corrections are dominant and can be precisely calculated. Uncertainties of the weak correction are mainly a result of uncertainties of the masses m_{top} and m_{Higgs} of the top quark and Higgs particles (Veltman 1977, 1980; Marciano & Sirlin 1984).

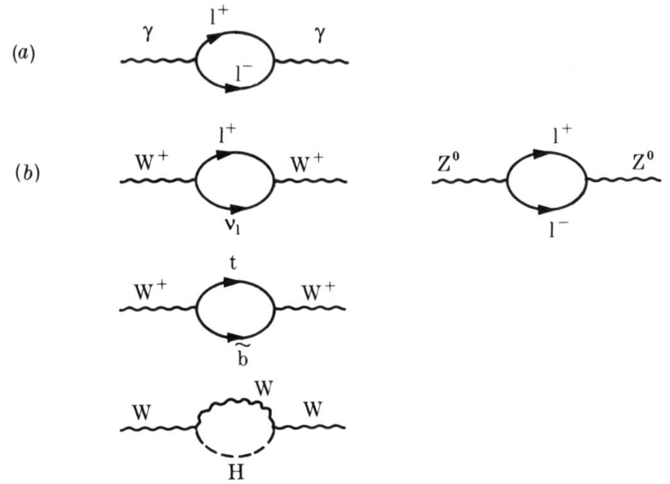

FIGURE 2. (*a*) Example of electromagnetic radiative correction. (*b*) Examples of weak radiative corrections.

Quantum chromodynamics is a similar gauge theory, except for the following points.

(i) The 'strong charge' is colour and each quark has one of three colour charges. However, the hadrons that are observable (so far) are colourless.

(ii) Interactions occur via the exchange of one of an octet of gluons (again spin-1 bosons but with mass $m_g = 0$), which carry colour charge and allow colour transitions according to SU(3).

(iii) Leptons are of course colourless, and so do not couple to the exchange gluons.

(iv) Because the gauge theory is non-Abelian, gluons self-couple (see figure 3), though not to other IVBS.

After the initial development of the q.c.d. theory, Altarelli & Parisi (1977) interpreted the lowest order QCD diagrams (figure 3) in terms of the parton model originally developed by Feynman (1969). In addition to validating the experimental successes of the parton model, a number of additional consequences that were not natural within the parton model were in accord with the data, in particular:

(i) the observation that only about 50 % of the nucleon momentum in eN or νN collisions could be attributed to quarks (the remainder is carried by gluons);

(ii) the observation that the strength of the strong interaction depended on the momentum transfer, Q, carried by the exchange particle. In fact,

$$\alpha_s = \frac{12\pi}{(33-2n_f)}\frac{1}{\ln(Q^2/\Lambda^2)}, \tag{2}$$

where n_f is the number of quark types, and Λ is a scale factor.

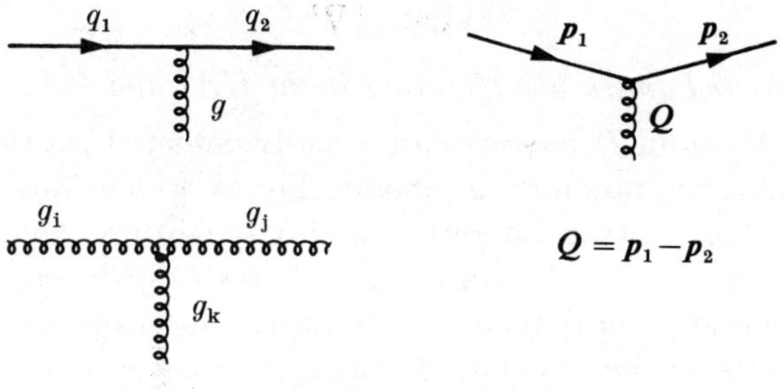

FIGURE 3. Lowest order QCD vertices. The momentum transfer, Q, transferred to the IVB is also illustrated in this figure.

In practice, because of the strength of the strong coupling and the importance of self-coupling diagrams, higher-order QCD corrections are important and represent a major computational challenge. Perturbative evaluations are appropriate only at values of Q^2 where divergences do not occur. In particular, in addition to the $p\bar{p}$ production diagrams for $W^{\pm}$ and Z^0 production shown in figure 4 (and described later), many additional higher-order diagrams involving a real or virtual gluon coupling, and gluon self-coupling, must be taken into account.

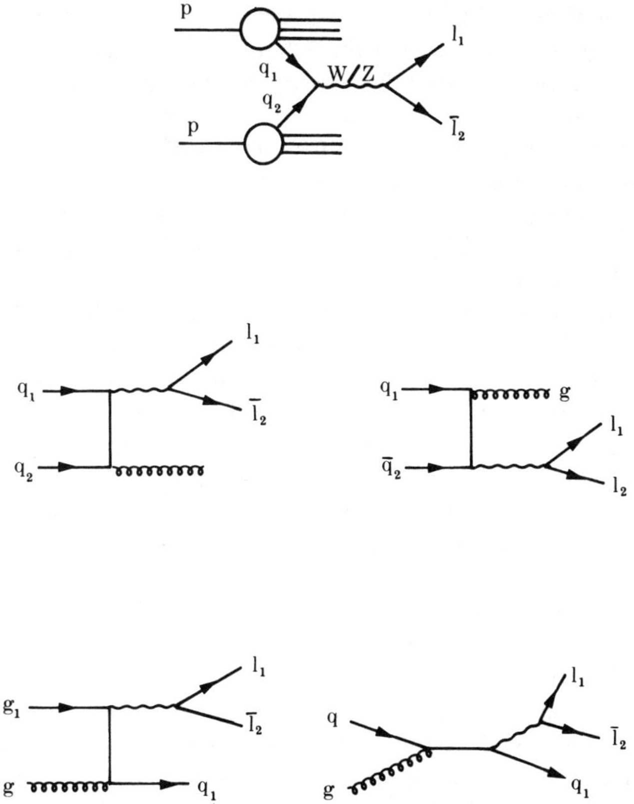

FIGURE 4. Lowest order diagrams for the hadronic production of $W^{\pm}$ and Z^0. (*a*) The Drell–Yan process. (*b*) The QCD annihilation and Compton processes.

3. Data

(a) *Lepton and quark identification in the UA1 and UA2 detectors*

At the collider, $W^{\pm}$ and Z^0 particles are usually detected via the electron decay mode (and, for UA1, the muon decay mode). For $W^{\pm}$ identification, this implies the need to detect an electron (muon) of high momentum. The $W \rightarrow e\nu$ decay is a distinctive topology in which (contrary to known QCD processes) a neutrino, which does not interact in the detector, is produced opposite to the electron. Hence, azimuthally opposite to the electron or muon, the existence of a non-interacting neutrino should be identifiable. For Z^0 identification, two electrons or muons with invariant mass M^{ll} characteristic of the Z^0 particle should be measured. Production is mainly via the Drell–Yan (1970) process, where constituent quarks of the colliding p and $\bar{p}$ interact at a weak-interaction vertex to produce a $W^{\pm}$ or Z^0 (figure 4*a*). However, additional processes involving the sea-quark or gluon content of the p and $\bar{p}$ contribute (typical examples are shown in figure 4*b*). Generally, strong-interaction processes involving the associated production of a quark or gluon result. Such quarks and gluons physically manifest themselves as hadronic jets. The detectors should identify $W^{\pm}$ or Z^0 production in association with jet activity.

The UA1 detector (Arnison *et al.* 1978) provides particle detection and associated energy measurement (including the momentum measurement of charged particles) at all azimuth angles ϕ, and to within 0.2° in polar angle θ of the beam line. Regions of limited detection capability exist in angles $\Delta\phi \approx \pm 15°$ vertically above and below the beam line. The UA2 detector (Banner *et al.* 1978) provides full azimuthal coverage in the range ($20° < \theta < 160°$) with respect to the beam line, with charged-particle momentum measurement in the regions ($20° < \theta < 40°$) and ($140° < \theta < 160°$). Details of the performance of each detector are given elsewhere (Timmer 1983; Mansoulié 1983; and references therein). However, some salient features should be noted.

(i) Electrons are measured with energy resolution $\Delta E \approx 0.15\sqrt{E}$ (E measured in giga-electronvolts), but with a systematic energy-scale uncertainty of *ca.* $\pm 3.0\%$ ($\pm 1.5\%$) in the UA1 (UA2) detectors. This systematic uncertainty limits our ability to compare the $W^{\pm}$ and Z^0 masses with predicted values. Background contributions to the electron signal result from jets which fake the electron identification criteria occasionally (typically 1 jet in *ca.* 5×10^4, from photon conversions or from the overlap of a high-energy photon with a π^+ or π^-).

(ii) Muons (UA1) are momentum measured with a resolution $\Delta p \approx 0.005\, p^2$ (p measured in GeV c^{-1}). The muon measurements are crucial to tests of the GSW model in different fermion doublets, but in the $W^{\pm}$ and Z^0 decay channels the measurement accuracy is limited.

(iii) For an ideal detector, a measurement of the imbalance in ϕ of detected transverse energy is possible, because momentum conservation requires $\Sigma_i \boldsymbol{p}_T^i = 0$, i extending over all detected particles, and subscript T representing the momentum component transverse to the beam axis. This imbalance is called 'missing transverse energy', or E_T^m. Depending on the interpreted physics process, it may be ascribed to the existence of a neutrino(s), or some other non-interacting particle(s). In practice, neither detector is perfect. By rejecting events with tracks pointing to ϕ regions of poor detection capability, UA1 measure $\Delta E_T^m \approx 0.7\sqrt{E_T(\text{tot})}$ where $E_T(\text{tot})$ is the total detected transverse energy. The resolution is measured to be Gaussian at the 4σ level. Because of the lack of particle detection at less than 20° to the beam line, UA2 has a slightly worse resolution, but more importantly, a non-Gaussian tail to the resolution function.

(iv) Jets can be identified in both detectors, with an energy resolution $\Delta E^J/E^J \approx 0.7(E^J)^{-\frac{1}{2}}$ for UA1, and $\Delta E^J/E^J \approx 0.32(E^J)^{-\frac{1}{4}}$ for UA2.

(*b*) *The data sample*

Figure 5 shows the distribution of p_T^e against E_T^m for all identified electrons in the standard UA2 analysis for $p_T^e > 15$ GeV c^{-1}. A significant population with $p_T^e \approx E_T^m \geqslant$ *ca.* 25 GeV is evident, as expected from $W^{\pm}$-decay. Other contributions

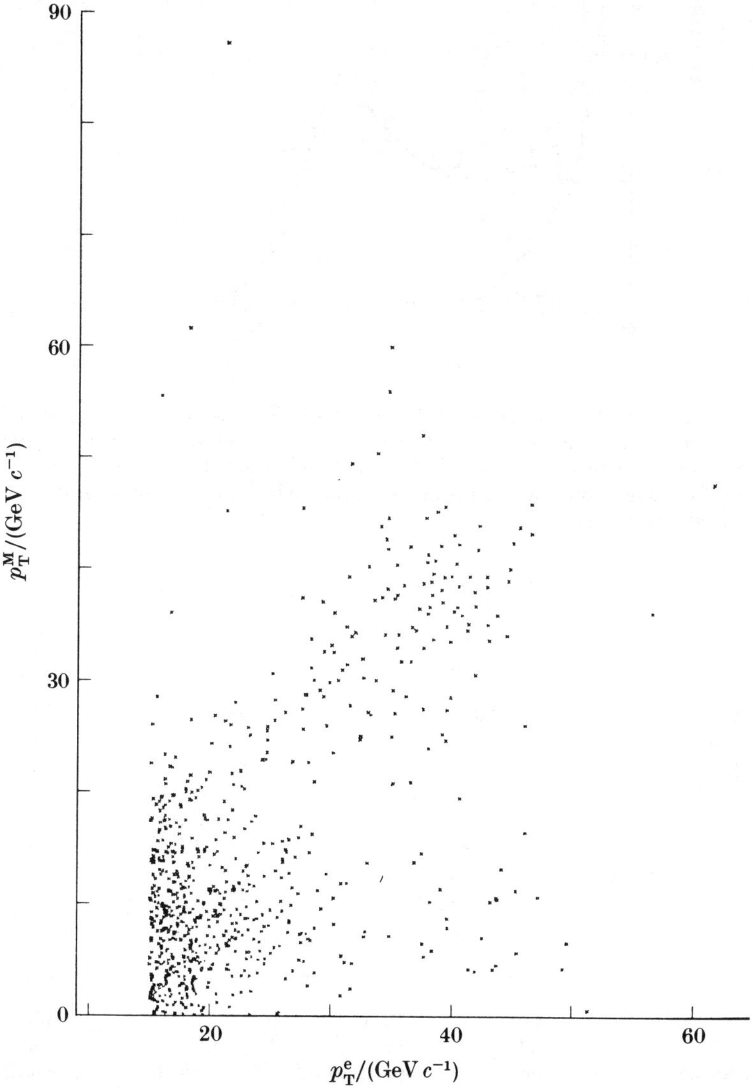

FIGURE 5. Distribution of electron p_T against missing p_T for all identified electrons of $p_T^e > 15$ GeV c^{-1} (UA2 data, 1983 and 1984; 840 electrons).

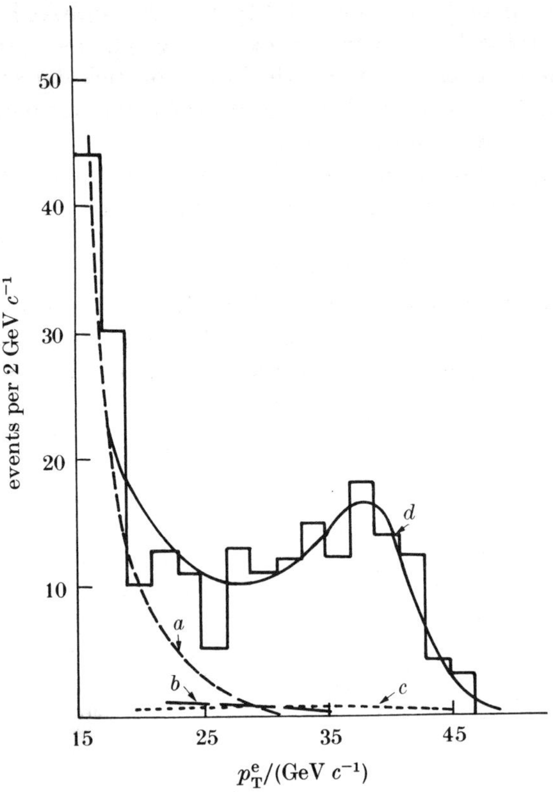

FIGURE 6. Distribution of p_T^e (UA2 data, 1983 and 1984) for electrons with $p_T^e > 15$ GeV c^{-1} satisfying topology selections for $W^\pm$ identification; 227 events are shown. Superimposed are expected contributions from (*a*) misidentified hadrons, (*b*) decays $W^\pm \rightarrow \tau^\pm \nu_\tau$, and (*c*) decays $Z^0 \rightarrow e^+e^-$, where one electron is not detected. The total expectation for $W^\pm \rightarrow e^\pm \nu_e$ decay is also shown as (*d*).

TABLE 2. UA1 AND UA2 EVENT SAMPLES

(*a*) $W^\pm \rightarrow e^\pm \nu_e$, $W^\pm \rightarrow \mu^\pm \nu_\mu$

	UA1	UA2	UA1 (μ)
$\int \mathscr{L}\, dt/(nb^{-1})$	390	452.5	362
p_T^l-cut/(GeV^{-2})	15	25	15
candidates	172	119	44
QCD background	5.3 ± 1.9	5.6 ± 1.7	2–3
$Z^0 \rightarrow l(l)$ one leg lost	—	4.4 ± 0.8	
$W^\pm \rightarrow \tau^\pm \nu_\tau$	9.1 ± 0.5	1.7 ± 0.2	not quoted
$W^\pm \rightarrow \tau^\pm \nu_\tau$; $\tau^\pm \rightarrow$ hadrons	2.7 ± 0.4	—	
signal	155 ± 13	107 ± 11	

(*b*) $Z^0 \rightarrow e^+e^-$, $Z^0 \rightarrow \mu^+\mu^-$

	UA1	UA2	UA1 (μ)
$\int \mathscr{L}\, dt/(nb^{-1})$	390	452.5	362
candidates	20	16	9
background (QCD)	small	less than 0.2 ($M^{ee} > 70$ GeV)	small
events used for mass determination	16	13	9

include Z^0-decay, electrons from the semi-leptonic decay of heavy quarks, electrons from the τ-decay of $W^{\pm}$, and a contribution of jets that are misidentified as electrons. After making topological cuts for $W^{\pm}$-decay, the UA1 and UA2 experiments identify about 300 leptonic decays of the $W^{\pm}$ (table 2) with a typical electron identification efficiency of *ca.* 75%. The p_T^e distribution of UA2 data is shown in figure 6, and the p_T^{μ} distribution of UA1 data is shown in figure 7. In figure 8, we superimpose a total of 45 $Z^0 \rightarrow e^+e^-$ and $Z^0 \rightarrow \mu^+\mu^-$ decays from UA1 and UA2; no other lepton pairs of $m^{ll} > 50$ GeV c^{-2} have been identified.

The collider data presented in this and following sections is restricted to that presented by the UA1 and UA2 collaborations before mid-June 1985. Specific

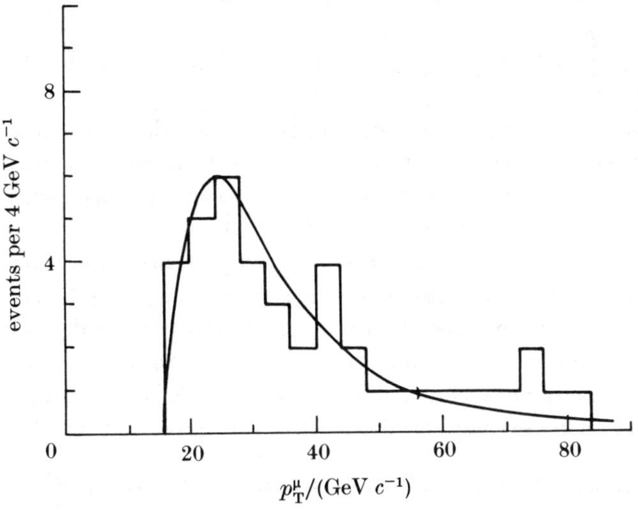

FIGURE 7. Distribution of p_T^{μ} (UA1 data, 1983 and 1984) for muons of $p_T^{\mu} > 15$ GeV c^{-1} identified as $W^{\pm}$ decays (44 events). The expectation for $W^{\pm} \rightarrow \mu^{\pm}\nu_{\mu}$ decay is superimposed, after taking into account the measured muon momentum resolution.

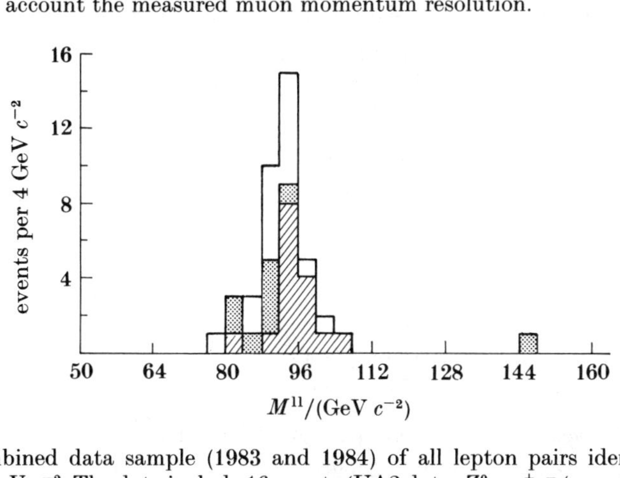

FIGURE 8. Combined data sample (1983 and 1984) of all lepton pairs identified with mass $M^{ll} > 50$ GeV c^{-2}. The data include 16 events (UA2 data: $Z^0 \rightarrow e^+e^-$ (open blocks), 20 events (UA1 data: $Z^0 \rightarrow e^+e^-$, 4 events without assigned mass value (hatched), and 9 events (UA1 data: $Z^0 \rightarrow \mu^+\mu^-$ (dotted). The highest mass ($\mu^+\mu^-$) pair is compatible within errors with Z^0-decay.

reference is made when data are available in written form. Other data are most recently described by Livan (1985) and Froidevaux (1985). The $W^{\pm}$ and Z^0 data samples for data collected at $\sqrt{s} = 546$ GeV (before 1984) are described by Banner *et al.* (1983), Bagnaia *et al.* (1983, 1984*a*, *b*) and Arnison *et al.* (1983*a*, *b*, *c*, 1984*a*, *b*, *c*, *d*), and are reviewed by Wahl (1984).

4. $W^{\pm}$ and Z^0 production

(*a*) *The $W^{\pm}$ and Z^0 production cross sections*

As shown earlier (figures 4*a*, *b*), both strong and electroweak processes contribute to the $W^{\pm}$ and Z^0 production cross section. Considering the Drell–Yan process of figure 4*a*, which contributes *ca.* 65% of the cross section:

$$\frac{d\sigma}{ds} = \sum_{a,\,b} \int dx_a \int dx_b \, \delta(x_a x_b s - \hat{s}) f_{a/A}(x_a) f_{b/B}(x_b) \, \hat{\sigma}(a+b \rightarrow W), \tag{4}$$

where x_a is the fractional longitudinal momentum of quark a in nucleon A; $f_{a/A}(x_A)$ is the probability that quark a in A carries momentum fraction x_a; and $\hat{\sigma}$ is the cross section for $W^{\pm}$ or Z^0 production, at effective centre of mass energy $\hat{s} = x_a x_b s$, where s is the nucleon–antinucleon centre of mass energy.

Higher order graphs, such as those shown in figure 4, are numerous and require detailed QCD evaluations. The nucleon quark contents $f_{a/A}(x_a)$ are taken from low-energy νN measurements (see the review of Savin 1984 and references cited) and evolved to the kinematic region of the collider, assuming the validity of QCD. Uncertainties of measured nucleon structure functions, and of the QCD momentum-transfer scale, dominate the theoretical uncertainties shown in table 3. Nevertheless

Table 3. $W^{\pm}$ and Z^0 production cross sections[a]

Predictions shown are from Altarelli *et al.* (1984*a*)

(*a*) UA1

process	$\sqrt{s}$/(GeV)	data (pb)	predicted (pb)
$Z^0 \rightarrow e^+e^-$		$41 \pm 21 \pm 7$	
$Z^0 \rightarrow \mu^+\mu^-$	546	$100 \pm 50 \pm 15$	42^{+13}_{-6}
mean		$58 \pm 21 \pm 9$	
$W^{\pm} \rightarrow \mu^{\pm}\nu_e$		$530 \pm 80 \pm 170$	
$W^{\pm} \rightarrow \mu^{\pm}\nu_{\mu}$	546	$670 \pm 170 \pm 150$	360^{+110}_{-50}
mean		$567 \pm 72 \pm 77$	
R (mean)	546	0.10 ± 0.04	0.117[b]

(*b*) UA2

process	$\sqrt{s}$/(GeV)	data (pb)	predicted (pb)	data (predicted) for $\sqrt{s}$ ratio
$Z^0 \rightarrow e^+e^-$	546	$110 \pm 39 \pm 9$	42^{+13}_{-6}	0.47 ± 0.24
	630	$52 \pm 19 \pm 4$	51^{+16}_{-10}	(1.21)
$W^{\pm} \rightarrow e^{\pm}\nu_e$	546	$500 \pm 90 \pm 50$	360^{+110}_{-50}	1.06 ± 0.23
	630	$530 \pm 60 \pm 50$	460^{+140}_{-80}	(1.26)
R	546	0.22 ± 0.09	0.117[b]	
	630	0.098 ± 0.087	0.11[b]	

[a] $\sigma_Z \equiv \sigma(\tilde{p}p \rightarrow Z^0 + X)\, B(Z^0 \rightarrow l^+l^-)$; $\sigma_W \equiv \sigma(\tilde{p}p \rightarrow W^{\pm} + X)\, B(W^{\pm} \rightarrow l^{\pm}\nu_e)$; $R \equiv \sigma_Z/\sigma_W$.
[b] Ratio sensitive to structure function used.

the agreement of recent calculations (Altarelli 1984*a*, *b*) with experimental data is satisfactory. The UA2 observation of 8 Z^0 events at $\sqrt{s} = 540$ GeV resulted in a larger cross section than expected. This was presumably a statistical fluctuation. To improve significantly future comparisons, an improved knowledge of the beam–beam luminosity (currently $\pm 8\%$) is required, as well as a better theoretical evaluation of QCD corrections. Data at $\sqrt{s} = 2$ TeV from the future Fermilab collider will significantly constrain present uncertainties of the QCD evaluation. Recent estimates on the $\sqrt{s}$ dependence of the cross section are given by Altarelli *et al.* (1984*b*).

It should be noted that what are measured are the partial cross sections

$$\sigma(\bar{p}p \rightarrow W^{\pm} + X)\, B(W^{\pm} \rightarrow l^{\pm}\nu_l) \tag{5a}$$

and

$$\sigma(\bar{p}p \rightarrow Z^0 + X)\, B(Z^0 \rightarrow l^+l^-), \tag{5b}$$

where B denotes the branching fraction of the $W^{\pm}$ and Z^0 into leptons.

(*b*) *Transverse* $W^{\pm}$ *and* Z^0 *motion*

Because of the Fermi motion of quarks within the nucleon, a typical IVB transverse momentum $\langle p_T^{W,Z} \rangle \leqslant$ *ca.* 500 MeV c^{-1} would be expected. However, the

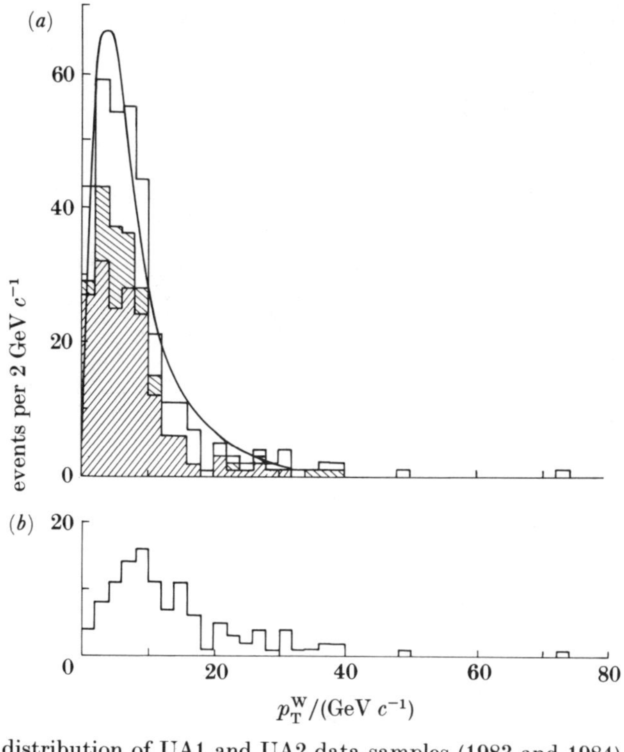

FIGURE 9. (*a*) p_T^W distribution of UA1 and UA2 data samples (1983 and 1984). Superimposed is the QCD expectation evaluated by Altarelli *et al.* (1984*a*) using DO1 (Duke & Owens 1984) structure functions and $\Lambda = 0.2$ GeV. Shading as figure 8. (*b*) p_T^W distribution of subsample of events having at least one associated jet of $E_T^J > 5$ GeV. All high-p_T Ws have associated jet activity. The QCD expectation is not shown because of differences in experimental acceptance of the detectors. UA1 and UA2 data.

data of UA1 and UA2 show $\langle p_T^{W,Z} \rangle \approx$ 7–9 GeV c^{-1}. This is in fact well predicted when QCD processes such as those in figure 4*b* are taken into account. The calculations of Altarelli *et al.* (1984*b*), shown in figure 9, are in good agreement with the data. As expected, the IVBs of highest p_T are associated with jet activity.

At $\sqrt{s} = 540$ GeV, the UA2 collaboration identified three events of very large p_T^W shown in figure 9 (Bagnaia *et al.* 1984*b*). At that time, the events were difficult to interpret as $W^{\pm}$ production resulting from known QCD processes. With increased statistics, additional events of that type have not been observed, and their production from QCD processes is the most likely explanation (Plothow-Besch 1985).

(*c*) *Associated jet activity*

Ellis *et al.* (1985) and Geer & Stirling (1985) have made perturbative-QCD calculations for one-jet and two-jet production associated to $W^{\pm}$ and Z^0. Their

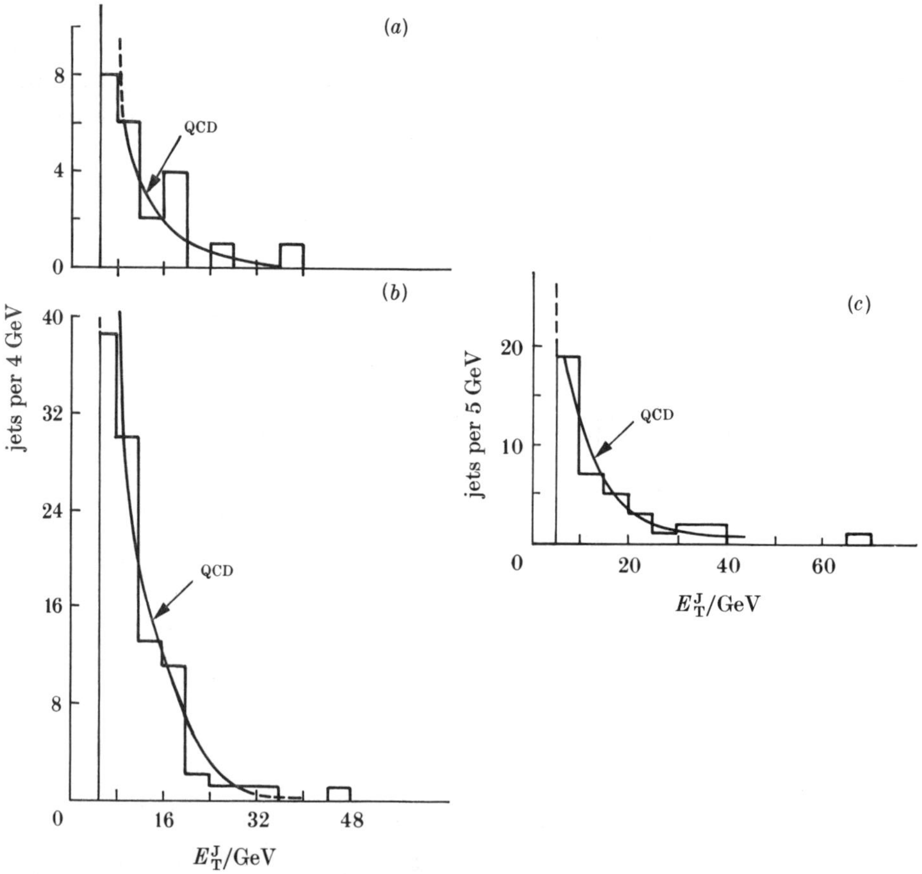

FIGURE 10. E_T^J distribution of identified jets having $E_T^J > 5$ GeV, associated to $W^{\pm}$ production. Superimposed QCD estimates use the calculations of Ellis *et al.* (1985). See experimental references for details of the QCD comparison. (*a*) UA1: $W^{\pm} \rightarrow \mu^{\pm}\nu$ (22 jets); (*b*) UA1: $W^{\pm} \rightarrow e^{\pm}\nu$ (89 jets); (*c*) UA2: $W^{\pm} \rightarrow e^{\pm}\nu$ (40 jets).

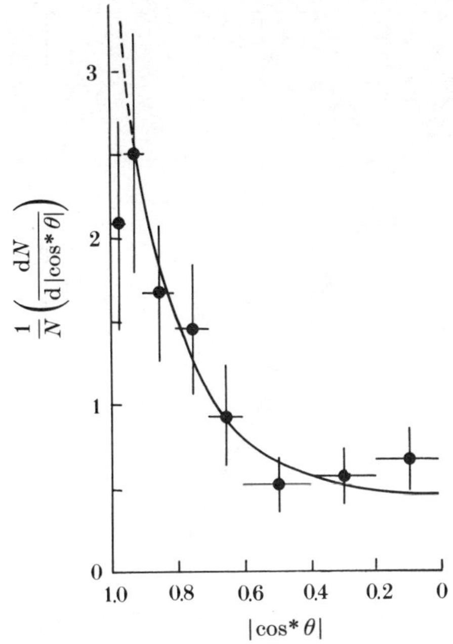

FIGURE 11. The $\cos\theta^*_J$ distribution for UA1 data ($p\bar{p}\rightarrow W+J+X$; 95 jets) in the W$^{\pm}$ jet centre of mass. The superimposed curve is that expected from gluon bremsstrahlung processes.

predictions for the jet E_T spectrum, dN/dE^J_T, are compared with data in figure 10. For $E^J_T \geqslant ca.$ 10 GeV, where perturbative calculations should be reliable, the agreement is good. The expected one-jet and two-jet rate is also in reasonable agreement with the data, for both UA1 and UA2 data. It should be noted that the predictions shown include the detector acceptances, but not a full treatment of the effect of fragmentation on the jet energy measured by the detectors.

In the W$^{\pm}$-jet centre of mass, figure 11 shows UA1 data (Geer 1985) for $dN/d\cos\theta^*$, where θ^* is the angle of the jet to the beam axis. The data are well described by the form $dN/d\cos\theta^* \propto (\sin\theta^*/2)^{-2}$, as expected for gluon bremsstrahlung.

To test quantitatively QCD expectations of associated jet activity, significantly more data having large E^J_T are required.

(*d*) *Longitudinal motion*

Because quarks a, b, involved in W$^{\pm}$ or Z^0 production do not carry the same fractional momentum, a net longitudinal motion of the produced IVB is expected, for example,

$$x_W = 2p^W_L/\sqrt{s} = x_a - x_b, \tag{6a}$$

$$m^2_W = x_a x_b s. \tag{6b}$$

The quantity x_Z is directly measured but the statistics are poor. Because no measurement of p^ν_L is available, x_W can be measured to within a quadratic

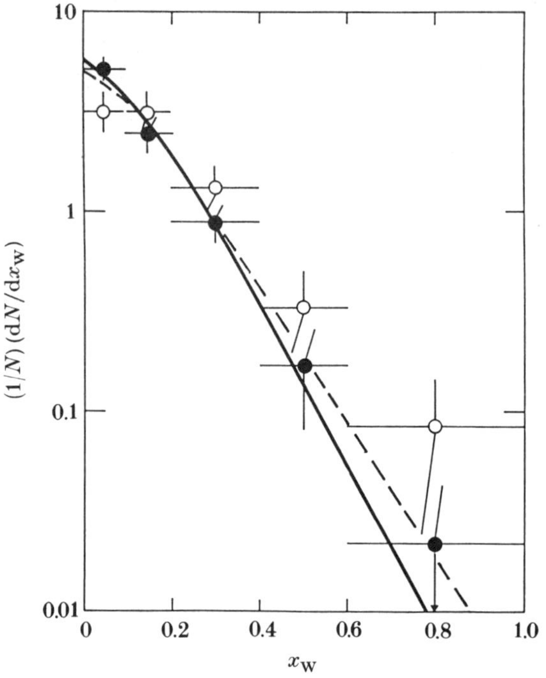

FIGURE 12. The x_W distribution for longitudinal $W^{\pm}$ motion at (○) $\sqrt{s} = 546$ GeV (59 events) and (●) $\sqrt{s} = 630$ GeV (113 events) (UA1 data). The superimposed expectations use structure-function parametrizations from Eichten *et al.* (1985); solid line shows $\sqrt{s} = 630$ GeV, broken line shows $\sqrt{s} = 546$ GeV.

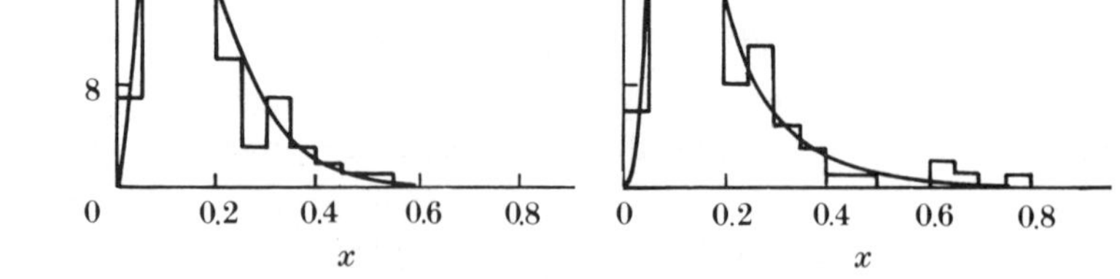

FIGURE 13. Longitudinal momentum fractions x_u, x_d carried by (*a*) u- and (*b*) d-quarks in the nucleon, assuming $W^{\pm}$ production as in figure 1*a*. Data shown are from UA1, 118 events, for which the electron charge is known at the 3σ-level. Superimposed expectations use the structure functions of Eichten *et al.* (1985).

ambiguity if m_W is known. In more than 80% of the examples, this ambiguity is resolvable. The UA1 data shown in figure 12 are compared with QCD expectations at $\sqrt{s} = 540$ GeV and $\sqrt{s} = 630$ GeV. Similar data exist from the UA2 experiment. By assuming $W^{\pm}$ production via the Drell–Yan process and with the use of a knowledge of the lepton charge, one can isolate $(x_u, x_{\bar{d}})$ for W^+ production, and $(x_{\bar{u}}, x_d)$ for W^- production. The UA1 data are compared with a QCD-evolved parametrization of structure function measurements (Eichten *et al.* 1985) in figure 13. Agreement is reasonable.

5. $W^{\pm}$ AND Z^0 DECAY

(*a*) *Lepton charge asymmetry in the decay* $W \rightarrow l\nu_l$

Because the production and decay vertices of figure 4 both involve $W^{\pm}$ exchange, and because $W^{\pm}$ exchange involves only left-handed fermion doublets (§2), the final-state lepton is constrained by angular momentum conservation to decay preferentially in the direction of the incident proton. The decay distribution takes the form

$$\frac{dN}{d\cos\theta^*} \propto (1+\cos\theta^*)^2, \qquad (7)$$

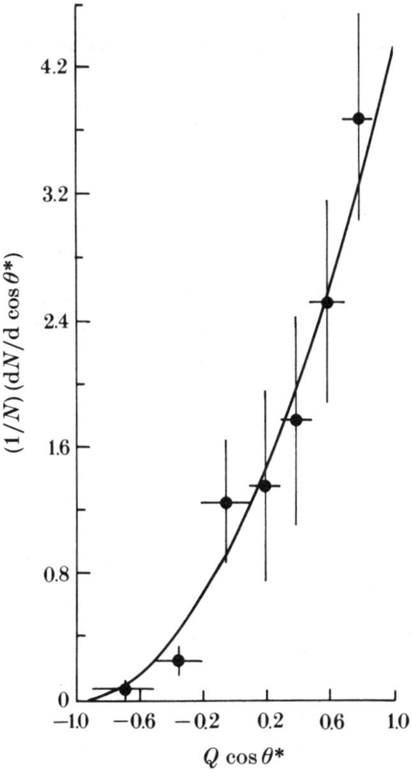

FIGURE 14. Lepton asymmetry (acceptance corrected) for the decay $W^{\pm} \rightarrow e^{\pm}\nu$ (UA1, 118 events). The superimposed curve is of the form $(1+\cos\theta^*)^2$, where $\cos\theta^*$ is defined in the text.

where θ^* is the angle of the lepton to the beam direction, in the $W^{\pm}$ rest frame. The resulting charge asymmetry is shown for UA1 data in figure 14, and agrees well with expectation. Data are available from UA2 in the angular range ($20° < \theta < 40°, 140° < \theta < 160°$). They measure a charge asymmetry $\alpha = (F-B)/(F+B)$, where F and B represent a decay lepton produced in the direction of the incident proton and antiproton, respectively. The UA2 collaboration measure $\alpha = 0.43 \pm 0.17$, to be compared with an expectation $\alpha = 0.52 \pm 0.06$ after taking sea-quark contributions and higher order processes into account.

Because two weak-interaction vertices are involved, pure left-handed and pure right-handed coupling (that is, V−A and V+A coupling) cannot be distinguished. If a third weak-interaction vertex is required, for example,

$$\bar{p}p \rightarrow W^+ + X; \quad W^+ \rightarrow \tau^+ \nu_\tau; \quad \tau^+ \rightarrow e^+ \nu_e \bar{\nu}_\tau, \tag{8}$$

the decay angular distribution of the e^+ differs for V−A and V+A coupling.

Because of the increasing effect of sea-quark contributions, the lepton asymmetry measured at the Fermilab collider ($\sqrt{s} = 2$ TeV) will be significantly reduced.

Because the Z^0-coupling is not pure left-handed, the expected asymmetry for Z^0 decay is small.

(b) *Fermion universality*

It is an assumption of the GSW model that (neglecting quark mixing) weak transitions occur within a fermion doublet, and that the coupling is independent of the fermion doublet involved. Independent of the doublet, the $W^{\pm}$ and Z^0 partial widths for a given fermion pair are

$$\Gamma_W(f_1, \bar{f}_2) = \frac{G_F m_W^3}{24\sqrt{2}\,\pi} C\phi, \tag{9a}$$

$$\Gamma_Z(f\bar{f}) = \frac{G_F m_Z^3}{6\sqrt{2}\,\pi\, C\phi'(g_V^2 + g_A^2)}, \tag{9b}$$

where C is the colour factor (1 for leptons, 3 for quarks), ϕ and ϕ' are kinematic phase-space factors suppressing the contribution of high-mass fermions (for example, the top quark), and g_V, g_A are defined in table 1. G_F is the Fermi coupling constant.

Because the UA1 experiment identifies muons, the expected equal coupling strengths of muons and electrons to IVBs (that is μ–e universality) can be tested. By using $\sqrt{s} = 540$ GeV, they obtain:

$$R_W^\mu = \frac{\sigma(\bar{p}p \rightarrow W^{\pm} + X;\ W \rightarrow \mu\nu_\mu)}{\sigma(\bar{p}p \rightarrow W^{\pm} + X;\ W \rightarrow e\nu_e)} = 1.26 \pm 0.4, \tag{10a}$$

$$R_Z^\mu = \frac{\sigma(\bar{p}p \rightarrow Z^0 + X;\ Z^0 \rightarrow \mu^+\mu^-)}{\sigma(\bar{p}p \rightarrow Z^0 + X;\ Z^0 \rightarrow e^+e^-)} = 2.40 \pm 1.70. \tag{10b}$$

The UA1 experiment has recently reported (Savoy-Navarro 1985) evidence for the decays $W^{\pm} \rightarrow \tau^{\pm}\nu_\tau$; $\tau^{\pm} \rightarrow \pi^{\pm}(n\pi^0)\,\nu_\tau$ and $\tau^{\pm} \rightarrow \pi^{\pm}\pi^{\mp}\pi^{\pm}(n\pi^0)\,\nu_\tau$. By using data at $\sqrt{s} = 540$ GeV, they quote limits

$$R_W^\tau = \frac{\sigma(\bar{p}p \rightarrow W^{\pm} + X;\ W^{\pm} \rightarrow \tau^{\pm}\nu_\tau)}{\sigma(\bar{p}p \rightarrow W^{\pm} + X;\ W^{\pm} \rightarrow e^{\pm}\nu_e)} = 0.96 \pm 0.45. \tag{11}$$

The statistical error on the above ratios should be significantly improved when the 1984 data of UA1 are released. In a different kinematic régime, μ–e universality is well tested, and τ–e universality is tested to an accuracy of less than *ca.* 15% via the reactions $e^+e^- \rightarrow \mu^+\mu^-$ and $e^+e^- \rightarrow \tau^+\tau^-$. (See the reviews of Pullia (1984) and Langacker (1984), together with new data of Fernandez *et al.* (1985) and Gan *et al.* (1985).)

Within the quark sector, results are less precise. The UA2 collaboration searches for the decays $W^\pm \rightarrow \text{jet} + \text{jet}$, and $Z^0 \rightarrow \text{jet} + \text{jet}$. These decay modes should account experimentally for about 60% of all $W^\pm$ and Z^0 decays. However, the resulting two-jet final state is topologically similar to that of more copiously produced QCD processes (a factor of *ca.* 75 at $\sqrt{s} = 540$ GeV), and must be observed as an enhancement of the jet–jet invariant mass m^{JJ} on the QCD continuum. By using data at $\sqrt{s} = 540$ GeV, they quote an enhancement of 100^{+80}_{-70} events, to be compared with an expectation of *ca.* 150 events (Weidberg 1984). To be convincing, more careful systematic studies, and increased statistics, are required.

The UA1 collaboration attempts to tag c, b, and t quarks via their semi-leptonic decay. They so far report twelve events compatible with the decay $W^+ \rightarrow t\bar{b}$ and its charge conjugate. Six events are published (Arnison *et al.* 1984*d*). However, from the GSW model, only two or three events are expected. The analysis is reported by J. D. Dowell (this symposium); further results are awaited with much interest.

From electroweak interference in the reactions $e^+e^- \rightarrow f^+f^-$, and from neutrino-induced neutral-current processes, the expected u- and d-quark couplings are well verified. The c- and b-quark couplings are determined to within *ca.* ±25%, and limited data on s-quark couplings are compatible with expectation (Pullia 1984; Langacker 1984; and references cited).

In summary, data from the CERN collider are in agreement with the assumption of fermion universality, but much improved limits should be a major priority of future collider experimentation.

(*c*) W$^\pm$ *and* Z^0 *masses, and the GSW model*

The Z^0 mass, m_Z, can be directly evaluated from the m^{ll} distribution of figure 8. However, because the neutrino longitudinal momentum p_L^ν is not measured, m_W is not known event by event. In practice, the $m_T^{e\nu}$ distribution, shown in figure 15 for UA2 data, is compared with model-dependent expectations. The most precise measurements so far are from the UA2 experiment:

$$m_Z = 92.5 \pm 1.3\,(\text{stat}) \pm 1.5\,(\text{syst})\ \text{GeV}\ c^{-2},$$

$$m_W = 81.1 \pm 1.1\,(\text{stat}) \pm 1.3\,(\text{syst})\ \text{GeV}\ c^{-2}.$$

Systematic uncertainties, resulting mainly from energy-calibration uncertainties in the measurement of electron energy (§3*a*) now dominate the error. In future collider experiments, calibration uncertainties should be significantly improved.

As noted in §2, all parameters of the GSW model are now defined, from measurements of α, G_F, m_W and m_Z. In practice, however, it is usual to evaluate

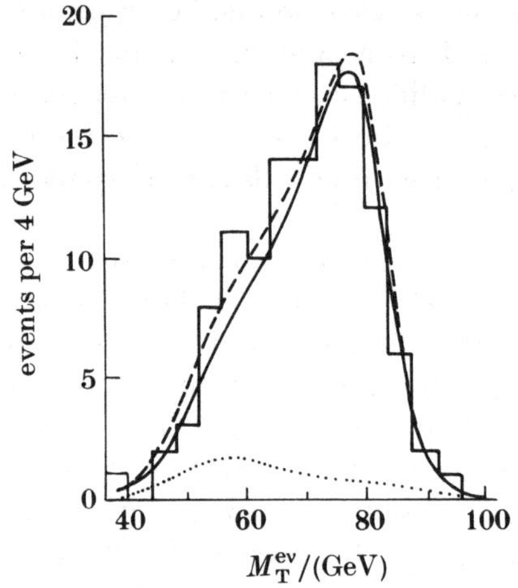

FIGURE 15. Transverse mass distribution (UA2 data: 128 events) for events satisfying $p_T^e > 17$ GeV c^{-1}, $p_T^\nu > 25$ GeV c^{-1}. The total expectation (broken line) includes contributions from $W^\pm \to e^\pm \nu_e$ decay (full line), together with other contributions (dotted line) which include misidentified hadrons, $W^\pm \to \tau^\pm \nu_\tau$ decay, and $Z^0 \to e^+e^-$ decay.

the quantities $\sin^2\theta_W$ and ρ. By using UA2 data alone, $\sin^2\theta_W$ can be defined, independently of the energy scale:

$$\sin^2\theta_W = 1 - m_W^2/m_Z^2$$
$$= 0.233 \pm 0.031. \qquad (12)$$

This definition, which assumes $\rho = 1$, is powerful in principle because experimental and theoretical uncertainties largely cancel, and the inclusion of radiative corrections is implicit. The measurement is at present, however, statistically limited.

Alternatively, $\sin^2\theta_W$ can be evaluated directly from the GSW model, after applying radiative corrections valid at the W mass, by:

$$\sin^2\theta_W(m_W) = \left(\frac{\pi\alpha}{\sqrt{2G_F}}\right)^2 \frac{1}{m_W^2}\frac{1}{(1-\Delta r)} = \frac{(37.2804)^2}{m_W^2}\frac{1}{(1-\Delta r)}, \qquad (13)$$

with
$$\rho^{\frac{1}{2}} = m_W/m_Z \cos\theta,$$

and m_W in units of GeV c^{-2} for numerical evaluations.

Here, Δr incorporates (dominantly) electromagnetic corrections, but also weak radiative corrections, and radiative corrections involving the Higgs boson. By using $m_{top} \approx 36$ GeV c^{-2} and $m_{Higgs} \approx 100$ GeV c^{-2}, Marciano & Sirlin (1984) estimate $\Delta r = 0.0696 \pm 0.0020$. With this estimate, $\sin^2\theta_W(m_W)$ and ρ can be evaluated:

$$\sin^2\theta_W(m_W) = 0.227 \pm 0.005 \pm 0.008,$$
$$\rho = 0.99 \pm 0.05 \pm 0.02.$$

This value can be compared with the world average (WA) (Marciano 1985) of existing non-collider data (after renormalization to the $W^\pm$ mass),

$$\sin^2\theta_W(\text{WA}) = 0.220 \pm 0.007.$$

By assuming $\rho = 1$, Δr can be evaluated for either UA2 data alone, or by combining $\sin^2\theta_W(\text{WA})$ and the measured values of m_W and m_Z:

$$\Delta r\,(\text{UA2}) = 0.092 \pm 0.101 \pm 0.029,$$
$$\Delta r\,(\text{WA}) = 0.048 \pm 0.034 \pm 0.030.$$

The main effect of Δr and ρ is to change m_Z and m_W, (differently), and consequently $(m_Z - m_W)$, for which the systematic uncertainty arising from energy calibration is small. Figure 16 compares the status of UA2 measurements with expectations of the GSW model, and with $\sin^2\theta_W(\text{WA})$. Even with an expected improvement (*ca.* $2^{-\frac{1}{2}}$), when new UA1 data are available, the existing measurement of Δr is barely significant.

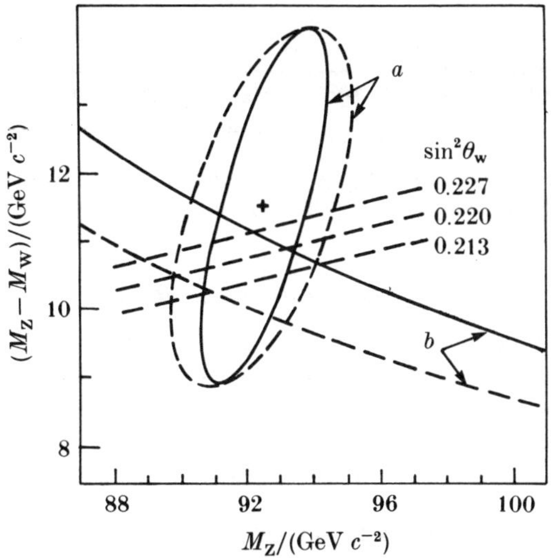

FIGURE 16. Distribution of m_Z against $(m_Z - m_W)$ for UA2 data. (*a*) Statistical (——) and systematic (----) uncertainties of the UA2 measurements. (*b*) Expectations of the G.S.W. model with (——) and without (----) radiative corrections as described in text. Also shown is the expectation based on $\sin^2\theta_W$ measurements from other experiments (see text).

To isolate a significant weak renormalization contribution to Δr, we require a measurement precision of less than *ca.* 0.003. Such measurements must await LEP. By 1989, m_Z will be measured to better than ± 50 MeV at LEP, and collider data should provide a measurement of $\delta(m_Z - m_W) \leqslant$ *ca.* 500 MeV c^{-2}. Furthermore, the CHARM 2 experiment (Busi *et al.* 1983; Panman 1985) is expected, via the processes $\nu_\mu e \rightarrow \nu_\mu e$ and $\bar{\nu}_\mu e \rightarrow \bar{\nu}_\mu e$ to measure $\sin^2\theta_W(m_W)$ to better than ± 0.005, and independently to measure ρ at $W^{\pm}$ momentum transfers of $|Q^2| \sim 10^{-2}\ (\text{GeV } c^{-1})^2$.

In the above superficial presentation, we ignored many accurate experiments that have measured $\sin^2\theta_W(m_W)$ and, in some experiments, ρ, over a large kinematic range. Davier (1984) gives a concise summary of the data. To illustrate one aspect of its importance, figure 17 shows an approximately constant value of $\sin^2\theta_W(m_W)$ over 16 orders of magnitude in $|Q^2|$ (from Langacker (1984) and Panman (1985).

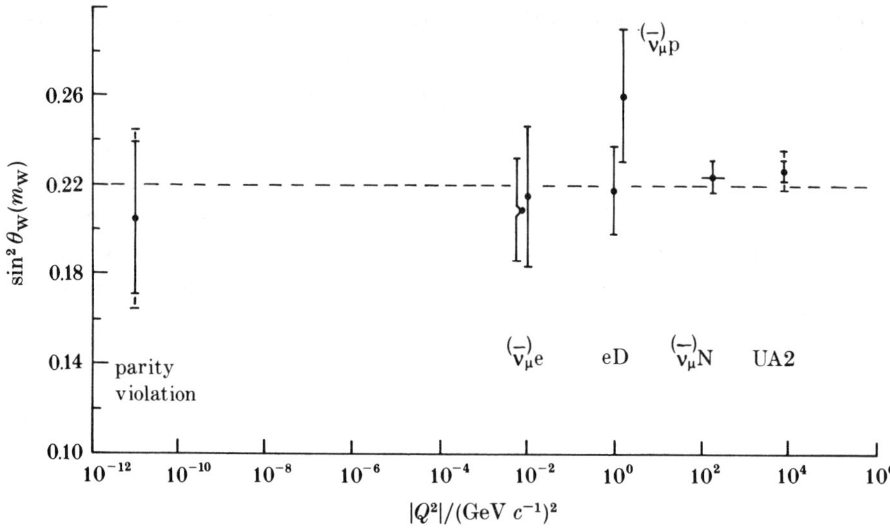

FIGURE 17. Measurements of $\sin^2\theta_W(m_W)$ at different IVB momentum transfers (from Langacker (1984) and Panman (1985)).

(d) Γ_Z, Γ_W, and additional neutrinos

After §5b, the expected total width for the decay of Z^0 and $W^{\pm}$ is obtained by summing the partial widths over all kinematically accessible decay states. By assuming three lepton and three quark doublets, and $m_{top} = 36$ GeV c^{-2},

$$\Gamma_W = 2.73 \text{ GeV } c^{-2},$$
$$\Gamma_Z = 2.74 \text{ GeV } c^{-2}.$$

If kinematically allowed, decays to additional lepton or quark doublets would increase Γ_W or Γ_Z. In particular, Z^0-decay into any new light neutrino ($Z^0 \rightarrow \nu_l \bar{\nu}_l$) would increase Γ_Z:

$$\Gamma_Z(\text{meas}) = \Gamma_Z(\text{G.W.S.}) + (n_\nu - 3)\,\Delta\Gamma_\nu; \quad \Delta\Gamma_\nu \approx 0.178 \text{ GeV } c^{-2}. \tag{14}$$

Experimentally, Γ_W is poorly measured because the mass m_W is not fully reconstructed; any limit results from a fit of known electron contributions to the $m_T^{e\nu}$ spectrum of figure 15. The best limit so far is:

$$\Gamma_W < 7 \text{ GeV } c^{-2} \text{ (at 90\% confidence level).}$$

Γ_Z can be directly measured (method 1), or can be evaluated from the cross-section ratio (method 2):

$$R = \frac{\sigma(Z^0 \rightarrow e^+e^-)}{\sigma(W \rightarrow e\nu)} = \frac{\sigma(Z^0 \rightarrow \text{all})}{\sigma(W \rightarrow \text{all})}\,\frac{\Gamma(Z^0 \rightarrow e^+e^-)}{\Gamma_Z}\,\frac{\Gamma_W}{\Gamma(W \rightarrow e\nu)}. \tag{15}$$

The ratio R is measured experimentally, and the ratio $\sigma(Z^0 \rightarrow \text{all})/\sigma(W \rightarrow \text{all})$ is evaluated by using QCD. By taking Γ_W, $\Gamma(W \rightarrow e\nu)$ and $\Gamma(Z^0 \rightarrow e^+e^-)$ from the GSW model, Γ_Z can be determined. The main advantage of this method is that most detector-dependent systematic uncertainties cancel, meaning that data from different experiments, and different decay channels, can be directly compared. The quoted UA2 results are:

$$\Gamma_Z(\text{method 1}) = 1.1^{+2.5}_{-0.8}\ \text{GeV}\ c^{-2}\ \text{(less than 4.6 GeV 90\% confidence level)}$$

$$\Delta n_\nu(\text{method 1}) \leqslant ca.\ 10\ \text{(90\% confidence level)}$$

$$\Gamma_Z(\text{method 2}) = 2.6^{+0.8}_{-0.5}\ \text{GeV}\ c^{-2}\ \text{(less than 3.7 GeV 90\% confidence level)}$$

$$\Delta n_\nu(\text{method 2}) \leqslant ca.\ 5\ \text{(90\% confidence level).}$$

(*e*) $Z^0 \rightarrow e^+e^-\gamma$

During the 1983 collider run, two events compatible with the decay $Z^0 \rightarrow e^+e^-\gamma$ were reported (Bagnaia *et al.* 1984*c*; Arnison *et al.* 1984*c*), from a total of twelve $Z^0 \rightarrow e^+e^-$ events. An additional event compatible with the decay $Z^0 \rightarrow \mu^+\mu^-\gamma$ was also reported, but because of the poorer mass resolution, the decay $Z^0 \rightarrow \mu^+\mu^-$ could not be excluded. In the 1984 collider run, no further events were reported. Given the kinematic configuration of the events, the probability that the two $(e^+e^-\gamma)$ events result from internal bremsstrahlung is *ca.* 4%; significantly more likely than other interpretation proposed (see the review of Wahl 1984).

6. Conclusion

With the measurement of m_W and m_Z at values predicted by the GSW model, and with the measurement of $\rho = 1$, evidence for the validity of the GSW model is overwhelming, at least for the leptonic doublet (e, ν_e) and the quark doublet (u, d'). In future tests of the GSW model, the emphasis will be towards a better verification of the model for other lepton and quark doublets, and a measurement of higher order radiative corrections.

A consequence of the simple GSW model is the existence of a single scalar particle, of unknown mass, but with a defined coupling to fermions and the weak IVBs. The observation of this particle, the Higgs scalar, is an important goal. Unfortunately, observation of the Higgs at the $p\bar{p}$ collider is unlikely because of its small production cross section. The production cross section increases with $\sqrt{s}$, and its observation is more likely at the proposed super-colliders ($\sqrt{s} \geqslant ca.$ 10 TeV).

A second consequence of all reasonable gauge theories incorporating the $W^{\pm}$ and Z^0 is the observation of the self-coupling of these particles at the predicted rate (for example, $Z^0 \rightarrow W^+W^-$). Again, because of the small cross sections involved, the observation of IVB self-coupling is most likely at the Fermilab collider ($\sqrt{s} = 2$ TeV) or at LEP II, when energies required for the direct production of (W^+W^-) pairs will be achieved.

In comparing $W^{\pm}$ and Z^0 production characteristics with the combined predictions of the GSW model and the QCD gauge theory for strong interactions, reasonable agreement is obtained. With increased statistics, these data will place powerful constraints on uncertainties inherent to the evaluation of higher order QCD processes.

I am grateful to the organizers of this meeting, Professors R. H. Dalitz and P. I. P. Kalmus, for the very productive atmosphere created.

I thank my colleages in UA1 and UA2 for many useful discussions, and for providing some material before publication. In particular, thanks are due to K. Eggert, A. Leveque and M.-N. Minard from the UA1 collaboration, and P. Darriulat, D. Froidevaux and L. di Lella from the UA2 collaboration.

The rapid transition from my manuscript and hand-drawn graphs to a final text is due to the CERN text processing unit and to E. Vial of the CERN drawing office, and is gratefully acknowledged.

References

Altarelli, G. & Parisi, G. 1977 *Nucl. Phys.* B **126**, 298–318.
Altarelli, G., Ellis, R. K., Greco, M. & Martinelli, G. 1984*a* *Nucl. Phys.* B **246**, 12–44.
Altarelli, G., Ellis, R. K. & Martinelli, G. 1984*b* *Z. Phys.* C **27**, 617–632.
Arnison, G. *et al.* 1978 A 4π solid-angle detector with the SPS used as a proton–antiproton collider at $\sqrt{s} = 540$ GeV. *CERN/SPSC/78-6.*
Arnison, G. *et al.* 1983*a* *Physics Lett.* B **122**, 103–116.
Arnison, G. *et al.* 1983*b* *Physics Lett.* B **126**, 398–406.
Arnison, G. *et al.* 1983*c* *Physics Lett.* B **129**, 273–282.
Arnison, G. *et al.* 1984*a* *Physics Lett.* B **134**, 469–476.
Arnison, G. *et al.* 1984*b* *Physics Lett.* B **135**, 250–254.
Arnison, G. *et al.* 1984*c* *Physics Lett.* B **147**, 241–248.
Arnison, G. *et al.* 1984*d* *Physics Lett.* B **147**, 493–508.
Bagnaia, P. *et al.* 1983 *Physics Lett.* B **129**, 130–140.
Bagnaia, P. *et al.* 1984*a* *Z. Phys.* C **24**, 1–17.
Bagnaia, P. *et al.* 1984*b* *Physics Lett.* B **139**, 105–114.
Banner, M. *et al.* 1978 Proposed to study $\bar{p}p$ interactions at 540 GeV c.m. energy. *CERN/SPS/78-8.*
Banner, M. *et al.* 1983 *Physics Lett.* B **122**, 476–485.
Busi, L. *et al.* 1983*a* Proposal *CERN/SPSC/83-24.*
Busi, L. *et al.* 1983*b* Proposal *CERN/SPSC/83-37.*
Chau, L. 1983 *Physics Rep.* **95**, 1–94.
Darriulat, P. 1984 In *Proc. 1984 Scottish Universities Summer School in Physics, St Andrews*; also CERN-EP/84-152.
Davier, M. 1984 In *Proc. XXI int. conf. on high-energy physics, Leipzig* (ed. A. Meyer & E. Wiezcorek), pp. 238–239. Zeuthen: Akademie der Wissenschaften der D.D.R.
Di Lella, L. 1985 Jet production in hadronic collisions. *A. Rev. nucl. Sci; Particle Sci.* (Submitted.)
Drell, S. D. & Yan, T. 1970 *Phys. Rev. Lett.* **25**, 316–320.
Duke, D. W. & Owens, J. F. 1984 *Phys. Rev.* D **30**, 49–54.
Eichten, E. *et al.* 1984 *Rev. mod. Phys.* **56**, 579–707.
Ellis, S. D., Kleiss, P. & Stirling, W. J. 1985 *Physics Lett.* B **154**, 435–440.
Fernandez, E. *et al.* 1985 *Phys. Rev. Lett.* **54**, 1624–1627.
Feynman, R. P. 1969 In *Proc. 3rd int. conf. on high-energy collisions, Stony Brook* (ed. C. N. Yang), pp. 237–258. New York: Gordon & Breach.
Froidevaux, D. 1985 In *Proc. 5th int. conf. on physics in collisions, Autun, France.* (In the press.)
Gan, K. K. *et al.* 1985 *Physics Lett.* B **153**, 116–120.
Geer, S. 1985 In *Proc. int. symposium on physics of* $p\bar{p}$ *collisions, Tsukuba*, KEK rep. 85-5, pp. 163–177; also *CERN-EP/85-63.*
Geer, S. & Stirling, W. J. 1985 *Physics Lett.* B **152**, 373–379.
Glashow, S. L. 1961 *Nucl. Phys.* **22**, 579–588.
Hasert, F. J. *et al.* 1983 *Physics Lett.* B **46**, 138–140.

Hung, P. Q. & Sakurai, J. J. 1978 *Nucl. Phys.* B **143**, 81–113.
Kobayashi, M. & Maskawa, K. 1973 *Prog. theor. Phys.* **49**, 652–657.
Langacker, P. 1984 In *Proc. XXII int. conf. on high-energy physics, Lepizig* (ed. A. Meyer & E. Wieczorek), pp. 215–238. Zeuthen: Akademie der Wissenschaften der D.D.R.
Livan, M. 1985 In *Proc. XVI symposium on multiparticle dynamics, Kyriat-Aravim, Israel* (In the press.). Paris: Editions Frontières.
Mansoulié, B. 1983 In *Proc. 3rd Moriond workshop on* p̄p *physics, La Plagne* (ed. Tran Thanh Van, J.), pp. 609–620. Paris: Editions Frontières.
Marciano, W. J. & Sirlin, A. 1984 *Phys. Rev.* D **29**, 945–951.
Marciano, W. J. 1985 In *Proc. Aspen winter physics conf., Aspen* (In the press.). Also *BNL 36147*.
Naroska, B. 1983 In *Proc. 1983 int. symposium on lepton and photon interactions at high energies* (ed. D. G. Cassel & L. Kreineck), pp. 96–131. Cornell: Laboratory of Nuclear Studies.
Panman, J. 1985 In *Proc. Aspen winter physics conf., Aspen* (In the press.). Also *CERN-EP/85-35*.
Plothow-Besch, H. 1985 In *Proc 5th workshop on* pp̄ *collider physics, St Vincent, Italy* (ed. M. Greco), pp. 57–86. Also *CERN-EP/85-86*.
Politzer, H. D. 1974 *Phys. Rep.* **14**, 129–180.
Pullia, A. 1984 In *Fifty years of weak-interaction physics* (ed. A. Bertin, R. A. Ricci & A. Vitale), pp. 333–415. Bologna: Italian Physical Society.
Ross, D. & Veltman, M. 1975 *Nucl. Phys.* B **95**, 135–147.
Salam, A. 1968 In *Proc. 8th Nobel symposium, Aspenäsgården*, p. 367. Stockholm: Almqvist & Wicksell.
Savin, I. 1984 In *Proc. XXII int. conf. on high-energy physics, Leipzig* (ed. A. Meyer & E. Wieczorek), pp. 251–279. Zeuthen: Akademie der Wissenschaften der D.D.R.
Savoy-Navarro, A. 1985 In *Proc. 5th workshop on* pp̄ *collider physics, St Vincent, Italy* (ed. M. Greco), pp. 196–211. Also *CERN-EP/85-64*.
Stone, S. 1983 In *Proc. 1983 int. symposium on lepton and photon interactions at high energies* (ed. D. G. Cassel & L. Kreinich), pp. 203–243. Cornell: Laboratory of Nuclear Studies.
Timmer, J. 1983 In *3rd Moriond workshop on* p̃p *physics, La Plagne* (ed. Tran Thanh Van, J.), pp. 593–608. Paris: Editions Frontières.
Veltman, M. 1977 *Nucl. Phys.* B **123**, 89–99.
Veltman, M. 1980 *Physics Lett.* B **91**, 95–98.
Wahl, H. D. 1984 In *Proc. XV int. symposium on multiparticle dynamics, Lund* (ed. G. Gustafso & C. Peterson), pp. 410–422. Singapore: World Scientific Publishing Co.
Weidberg, A. R. 1964 In *Proc. XV int. symposium on multiparticle dynamics, Lund* (ed. G. Gustafso & C. Peterson), pp. 349–365. Singapore: World Scientific Publishing Co.
Weinberg, S. 1967 *Phys. Rev. Lett.* **19**, 1264–1266.

Proc. R. Soc. Lond. A **404**, 213–232 (1986)
Printed in Great Britain

New physics from the CERN collider

BY J. D. DOWELL†
Department of Physics, The University of Birmingham, P.O. Box 363, Birmingham B15 2TT, U.K.

Experimental results from the CERN collider, having features that may not be explicable within the standard model of the strong, electromagnetic, and weak interactions, are described. They include events with dimuons and events with jets accompanied by missing energy. The evidence for the top quark is also presented.

1. INTRODUCTION

New physics, in the context of the so-called standard model of the strong and electroweak interactions, is generally considered to mean something that is not explicable within that model. Ellis (1984) discusses several indications, at the few-event level, of possible new phenomena of this type. With more data, some of these have since been found to be compatible with the standard model and are now interpreted as statistical fluctuations. However, two topics remain that could still imply 'new physics': events containing two muons, and events with missing energy. A less strict interpretation of 'new', on the other hand, might allow the inclusion of new observations that *are* expected within the model, and so I shall also discuss the evidence for the top quark.

At the present time, all the positive evidence for new things comes from the UA1 experiment (Astbury *et al.* 1978), though I shall mention UA2 (Banner *et al.* 1978) where appropriate. A few remarks are needed about the apparatus before discussing the physics. The UA1 detector (figure 1) consists of a segmented calorimeter surrounding a central tracking chamber in a uniform magnetic dipole field of 0.7 T and covering almost 4π sr. The calorimeter detects electromagnetically interacting particles (electrons and photons) by using a lead-scintillator sandwich, and strongly interacting particles (hadrons) by using this and a surrounding iron-scintillator device. Light pulses proportional to the energies of the particles, which are totally absorbed, are detected by photomultipliers and recorded in digital form on magnetic tape. The iron acts as the return yoke of the magnet and the detector is said to be hermetic, as all interacting particles at angles greater than 0.2° to the beam direction are recorded. Nevertheless, because high-energy particles are generally produced at small angles, about one-third of the energy of an event on average stays in the beam pipe. However, the detected 'transverse energy', $E_T = E \sin\theta$, where θ is the angle of a particle to the beam direction and E its energy, should balance if there are no non-interacting particles (for example, neutrinos) produced with appreciable transverse energy. Muons penetrate the calorimeters with minimum ionization loss, but are detected by further tracking

† Present address: E.P. Division, CERN, 1211 Geneva 23, Switzerland.

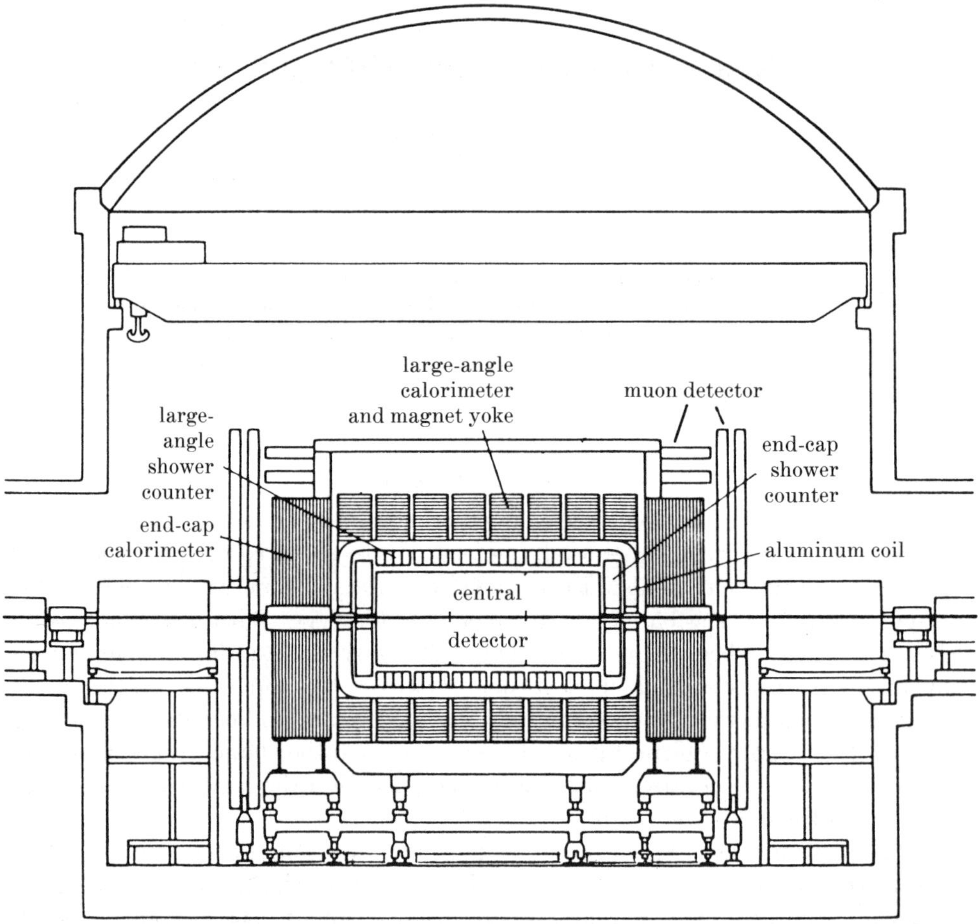

FIGURE 1. A view of the UA1 detector from the side showing the central track detector, electromagnetic calorimeters (shower counters), hadron calorimeter and magnet yoke, the magnet coil and the muon chambers. The electromagnetic calorimeters have four depth samplings and the hadron calorimeters two.

chambers outside the magnet yoke and can be taken into account in checking the transverse energy balance. The energy resolution of the calorimeters are given by

$$(\Delta E/E)_{\text{em}} = 0.15/\sqrt{E},$$

$$(\Delta E/E)_{\text{had}} = 0.80/\sqrt{E},$$

where E is in gigaelectronvolts (GeV). The momenta of charged particles can be determined from their curvature in the magnetic field with an accuracy

$$\Delta p/p \approx 0.005p,$$

where p is in GeV c^{-1}, if they are travelling at a sufficiently large angle to the field direction. Muons have to pass through about nine interaction lengths of material and are identified by the geometrical matching of the central detector and muon chamber tracks, to the limits of multiple scattering, together with their minimum ionization loss (figure 2). Backgrounds are principally a result of the decays of π and K mesons within the central detector.

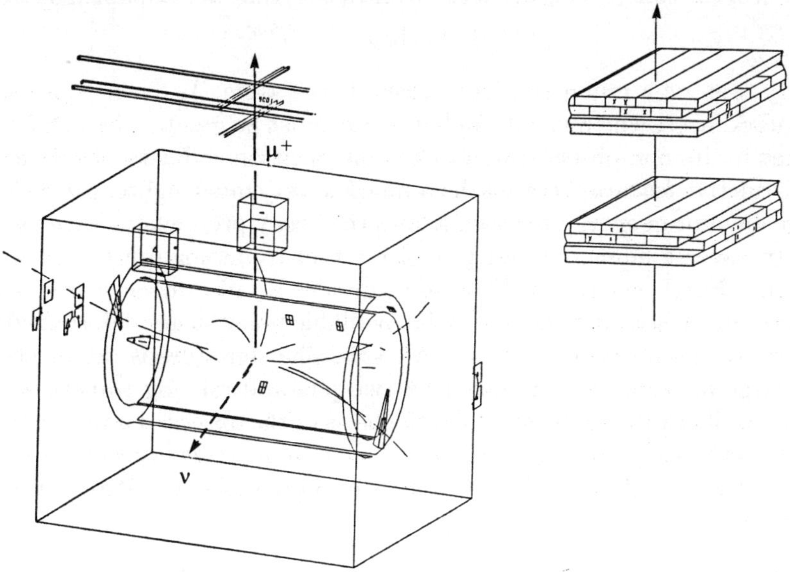

FIGURE 2. A $W \rightarrow \mu\nu$ decay illustrating the operation of the muon detector (drift tubes), the hadron calorimeter cells (boxes) and the missing transverse energy (neutrino).

As the phenomena of interest involve the primary interactions of quarks and antiquarks in the colliding beams, the events include 'jets' from quark fragmentation. In the UA1 detector, a jet is defined, for experimental purposes, as a localized cluster of energy (calorimeter cells) within a cone of semi-aperture $\Delta R \leqslant 1$, where $\Delta R = \surd[(\Delta\eta)^2 + (\Delta\phi)^2]$, and $\Delta\eta$, $\Delta\phi$ are respectively the differences in pseudo-rapidity and azimuthal angle of a cell direction from the jet-axis. Here η is defined as $-\ln \tan\frac{1}{2}\theta$, where θ is the angle of a particle to the beam direction, and ϕ is the azimuthal angle around the beam. To establish a new phenomenon requires the elimination not only of conventional explanations but also of spurious sources of background arising from the limitations of the detector. The latter is the primary task of the experimenter and relevant details of the techniques used will be discussed in the context of the three physics topics that rely on electron, muon, and missing transverse energy detection. UA2 differs from UA1 by having no muon detection and an incomplete solid angle coverage for its calorimeters. Also, its magnetic field extends over only a very restricted part of the apparatus. It is for these reasons that it is less sensitive to the phenomena to be discussed.

2. The top quark

Results from the 1983 operation of the collider yielded six events consistent with the decay of the top quark (Arnison *et al.* 1984*a*) and a mass estimate of $30 < m_t < 50$ GeV c^{-2}. In 1984 the total data sample was increased by a factor of three, so we shall discuss here the present status of the top quark evidence based on an integrated luminosity of 390 nb^{-1}.

According to the standard model the top quark should decay semi-leptonically about 12 % of the time, giving an electron and a neutrino, accompanied by a bottom quark,

$$t \rightarrow be\nu.$$

The decay rate via a muon and its neutrino is the same. As a consequence of the large t-quark mass, which is not predicted but must be greater than 22.5 GeV c^{-2} to account for its non-observation at electron–positron colliders, the decays have rather distinctive features because the b-quark is very much lighter (*ca.* 5 GeV c^{-2}). The large amount of energy released allows the decay products to have very large relative transverse momenta and provides a tool to distinguish the events from background. Furthermore, the W-boson provides a tailor-made source, as up to a quarter of its decays, depending on the available phase-space determined by the t-quark mass, should be $W^+ \rightarrow t\bar{b}$ (or $W^- \rightarrow \bar{t}b$. The signature is rather clear – an isolated large-p_t lepton accompanied by two jets and missing transverse energy (neutrino) as illustrated in figure 3. On the basis of the observed number of $W \rightarrow e\nu$ decays, a reliable estimate of the number of such events to be expected can be made for a given t mass. The principal source of background is the strong interaction (QCD) process,

$$q\bar{q} \rightarrow b\bar{b}g,$$

where g denotes a gluon. This would normally produce a three-jet event. However, the b or $\bar{b}$ may decay semi-leptonically:

$$b \rightarrow cl\bar{\nu};$$

yielding, in extreme examples, a configuration similar to that sought. This background cannot be removed and must therefore be estimated. However, more seriously (from an experimental point of view), 'fake' electrons are possible in one of the three jets, resulting from the overlap of a charged hadron and a photon which together conspire to simulate an electron-like signal in an electromagnetic calorimeter cell. It is necessary to use the full power of the UA1 apparatus to reduce this background to a low level.

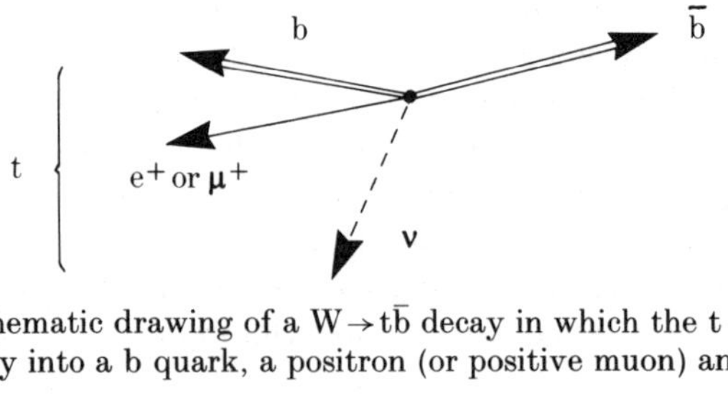

FIGURE 3. A schematic drawing of a $W \rightarrow t\bar{b}$ decay in which the t quark decays semi-leptonically into a b quark, a positron (or positive muon) and a neutrino.

In the 1983 data sample (Arnison *et al.* 1984*a*), there were six candidate events for top quarks; three having an electron and three a muon. Figure 4 shows an electron event. In the following we shall discuss the status of the combined 1983 and 1984 *electron* sample only, as the muon analysis is still in progress. The data

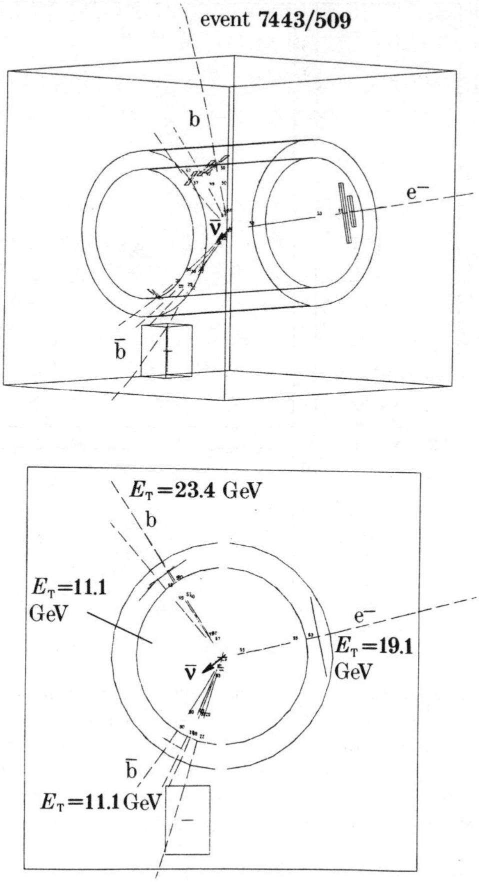

FIGURE 4. Two views of a $W \rightarrow \bar{t}b$ candidate. The $\bar{t}$ is assumed to decay into $\bar{b}e^-\bar{\nu}$.

yielded 417 candidates having a jet plus an electron with $E_T^e > 15$ GeV. These were reduced to 321 by removing γ converions in the beam pipe or tracking chambers, which are easily recognized because of the opposite bending of the electron and positron in the magnetic field. Next, an isolation criterion was applied to the electron (Σp_t (tracks) and ΣE_T (cells) are less than 1 GeV in a cone $\Delta R \leqslant 0.4$ around the electron), a x^2 cut on the energy deposition profile in the four depth samplings of the electromagnetic calorimeter, compared with that expected from test-beam electrons, and a requirement that the energy deposit in the hadron calorimeter be less than 0.2 GeV. Finally, matching in position and angle between the measured track and the struck calorimeter cell were required as well as agreement, within three standard deviations, between the track momentum and the calorimeter energy. The efficiency of these cuts for retaining electrons was tested on known electrons from $W \rightarrow e\nu$ decays.

The resulting sample contains nine events with an electron and two jets where E_T (jet 1) exceeds 8 GeV and E_T (jet 2) exceeds 7 GeV. Figure 5 is a mass plot of these, together with the three muon events from the 1983 data. The 'masses' are

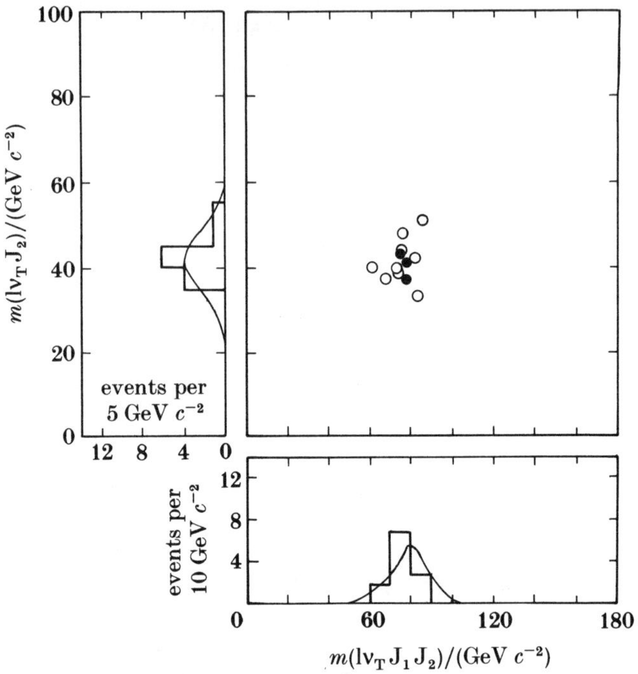

FIGURE 5. A two-dimensional mass plot of nine electron top candidates from the UA1 1983 and 1984 data (○) and three muon top candidates from the UA1 1983 data (●). The abscissa is the mass combination involving two jets ($E_T(J1) > 8$ GeV, $E_T(J2) > 7$ GeV), the lepton and the transverse energy of the neutrino. The ordinate is the mass including the lepton, neutrino and lower E_T jet, which should correspond to the top quark. The curves are the expectations for $W \to t\bar{b}$ decay with $m_t = 40$ GeV c^{-2}.

calculated by using the transverse energy of the neutrino (missing E_T), and so are the lower limits to the true masses. The lower E_T jet is assumed to come from t decay. If the events originate from $W \to t\bar{b}$ the mass combination including both jets should approximate to the W mass. The curves are what would be expected for a t mass of 40 GeV c^{-2}. It is seen that the observed events are, from this point of view, compatible with the $W \to t\bar{b}$ hypothesis.

Before concluding that the t quark has been observed, it is necessary to consider whether the background could simulate the signal and whether the observed number of events is indeed compatible with what would be expected. Without entering into the full details, it is possible to make a reliable estimate of the 'fake' electron background with the use of the actual data by relaxing certain of the conditions used to select electrons, and then estimating from the number of events so selected, how many would satisfy the electron criteria. Approximately one event is expected to be background from this analysis. Estimating the genuine prompt lepton background is more uncertain, but it can be done by comparing a theoretical QCD model with measured muon plus at least one-jet events to determine the relevant rates (cross sections). Again, approximately one event is expected.

The expected rate of $W \to t\bar{b}$ events of the type observed is calculated, as already discussed, starting from the $W \to e\nu$ events and allowing for branching fractions, acceptances, etc. Three events are expected, to be compared with the nine electron events (which include possibly two background events). Thus the observed signal appears too high to be accounted for entirely by $W \to t\bar{b}$. An alternative source is the process

$$p\bar{p} \to t\bar{t} + \ldots,$$

which is capable of yielding events with a similar configuration and comparable rates.

In summary, UA1 has observed a signal in lepton plus two jets plus missing transverse-energy events that is compatible with a mixture of $W \to t\bar{b}$ and $p\bar{p} \to t\bar{t}$ in which the t quark decays semi-leptonically. If this is the correct interpretation, the t mass is 40 ± 10 GeV c^{-2}. UA2 is more prone to fake electron background and less able to measure missing transverse energy. Its results do not conflict with this observation (Banner 1985).

3. Dimuon events

The UA1 detector is able to identify electrons and muons. Both of these can provide signatures for interesting phenomena. For example, heavy quarks (c, b, t) are expected to decay semi-leptonically 10–12 % of the time whereas the lighter quarks (u, d, s) do not give prompt leptons but lead to π and K mesons which then decay into leptons. Muons have two additional advantages over electrons: (*a*) they can be identified in jets, because the calorimeters absorb all strongly interacting particles; and (*b*) they can be distinguished from other particles at much lower values of transverse momentum (*ca.* 3 GeV c^{-1}) because of their penetrating property. Although π and K decays in flight can provide significant backgrounds in events with single muons, their contribution is very much lower in dimuon events, which are therefore a particularly fruitful source of interesting physics. Results from 1983 have already been published (Arnison *et al.* 1985). Here, the combined 1983 and 1984 data are discussed.

Dimuons may have unlike- or like-sign charges. There are three conventional sources of unlike-sign dimuons: (*a*) decays of heavy vector bosons (J/ψ, Υ, Z^0); (*b*) the Drell–Yan process; and (*c*) heavy quark decays. In the present paper we shall limit the discussion to $p_t > 3$ GeV c^{-1} for either muon or dimuon masses greater than 6 GeV c^{-2}, which eliminates the J/ψ. The Drell–Yan process (Kenyon 1982) is well established from experiments at lower energies and is the annihilation of a quark with an antiquark to give a muon pair by the electromagnetic interaction – i.e. via an intermediate virtual photon:

$$q\bar{q} \to \gamma^* \to \mu^+\mu^-.$$

Heavy quark–antiquark pairs, created in strong interactions and each decaying semi-leptonically, produce muon pairs about 1 % of the time. Here the muons will generally be inside or near to jets. However, as t quarks decay into b quarks and b quarks into c quarks a muon may be produced at a later stage in the decay chain. Furthermore, the sign of the muon alternates at each stage. Thus it is possible to obtain like-sign as well as unlike-sign pairs from heavy quark decays though the p_t cut preferentially suppresses the like-sign pairs because of their softer spectrum (daughter effect). The high mass of the t quark means that it has a low cross section and c quark detection is reduced by the p_t cut. Consequently, $b\bar{b}$ production is expected to be the dominant heavy flavour source of dimuons, for example:

$$p\bar{p} \rightarrow b\bar{b} + \ldots$$

$$\bar{b} \rightarrow \bar{c}\mu^{+}\nu$$

$$b \rightarrow c\mu^{-}\bar{\nu};$$

$$pp \rightarrow b\bar{b} + \ldots$$

$$\bar{b} \rightarrow \bar{c}\,\bar{d}\,u, \quad \bar{c} \rightarrow \bar{s}\mu^{-}\bar{\nu}$$

$$b \rightarrow c\mu^{-}\bar{\nu}.$$

Finally, $W \rightarrow t\bar{b}$ is a like-sign source but is expected to yield less than one event in the present data sample.

It is, of course, interesting to search for other sources that would represent new physics, such as the decays of new heavy leptons. However, it is first necessary to consider whether the observed events are amenable to conventional explanations. A phenomenon of special interest, because it has not previously been studied, that must also be taken into account is the prospect of $B^0-\overline{B^0}$ mixing. The b quark may combine with a $\bar{d}$ quark or an $\bar{s}$ quark to form a B^0_d or a B^0_s meson (with corresponding anti-mesons $\overline{B^0_d}$ and $\overline{B^0_s}$ containing $\bar{b}$ quarks). As the mesons are neutral, doubly weak transitions are possible which permit B^0 to change into $\overline{B^0}$ and vice versa. Analogous to the $K^0-\overline{K^0}$ system, the mass eigenstates are linear combinations of B^0 and $\overline{B^0}$. The mass difference between the two combinations determines the transition rate and several oscillations between B^0 and $\overline{B^0}$ are expected during the lifetime for the $B^0_s-\overline{B^0_s}$ system, the relevant parameter being $\Delta M/\Gamma$, where ΔM is the mass difference and Γ the decay width. The value of $\Delta M/\Gamma$ is estimated to be *ca.* 2.5 (Ali & Jarlskog 1984). The consequence is that like-sign and unlike-sign pairs are approximately equally likely from this source. However, charged B mesons do not contribute to this mechanism, and B^0_d mesons are expected to have small mixing, so the net effect of full mixing for the B^0_s situation is to raise the like : unlike sign ratio by about 0.2.

Experimentally, all candidates are carefully scanned to eliminate obvious decays in flight (kinks) and leakage through cracks between calorimeter cells. A total of 208 muon pairs result, 9 of which are Z^0 decays and 72 of which have like sign. The background is calculated in a relatively straightforward way by making use of the real data for events with one detected muon. The method is to calculate the probability that another particle in the event will decay (assuming 50 % pions and 25 % kaons) and then to apply all the experimental selection criteria to the resulting decay muon. A total background of 10.7 events is obtained and is shown in figure 6 together with the 199 low-mass events.

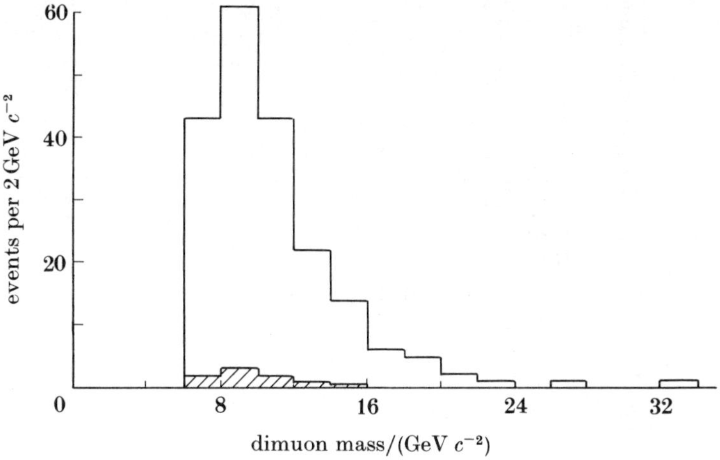

FIGURE 6. The dimuon mass spectrum for muons with $p_t > 3$ GeV c^{-1} and $m_{\mu\mu} > 6$ GeV c^{-2}. The shaded region is the calculated background (10.7 events). The 9 Z^0 events are not shown. Data are for a total of 199 events.

An obvious way to attempt to distinguish the sources of the events is by the isolation of the muon – i.e. whether it is in or near to a jet. For this purpose, isolation is defined as $\Sigma E_T \leqslant 4$ GeV in a cone $\Delta R \leqslant 0.7$ surrounding the muon. A scatter plot of the 199 events, excluding the Z^0 events, is given in figure 7, showing the transverse energy deposited in the cone around each muon and curves expected for 'minimum bias' (i.e. typical) events and heavy flavour (quark) production. After applying an isolation cut, the events may be classified into four categories: like- and unlike-sign, and isolated or non-isolated (figure 8). Note that an event is labelled non-isolated if either muon is in that class. The expected origins are summarized in table 1.

Obviously, a fraction of heavy quark decays can satisfy the isolation criterion (see figure 7).

The UA1 experimenters have made a preliminary attempt to account for the events in terms of the physical processes outlined above. By assuming that the 83 non-isolated, unlike-sign values originate entirely from heavy quark decays, the heavy quark contribution to the isolated, unlike-sign class (*ca.* 21 events) may be estimated. The remainder of this mass spectrum is then fitted to Υ decay (including excited states) and Drell–Yan processes, giving the following cross sections:

$$\sigma_{DY}(m > 11 \text{ GeV } c^{-2}) = 0.31 \pm 0.12 \text{ nb};$$

$$B\sigma(\Upsilon) = 0.34 \pm 0.21 \text{ nb},$$

where the errors are statistical and B is the branching ratio for $\Upsilon \rightarrow \mu^+\mu^-$. The Drell–Yan cross section agrees well with extrapolations from lower energies and

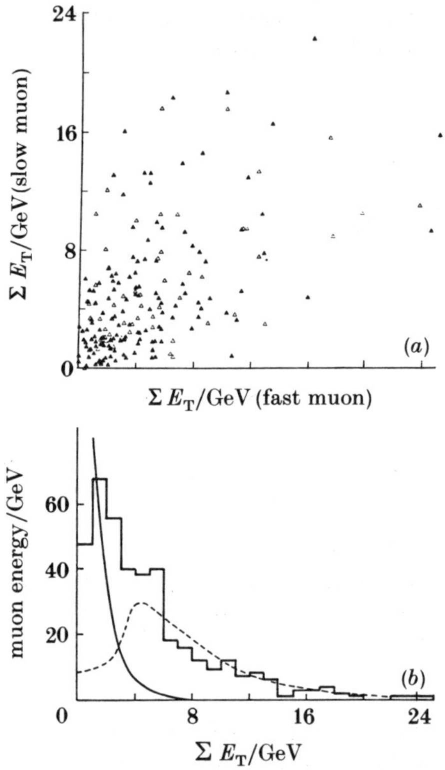

FIGURE 7. (*a*) A two-dimensional plot of the transverse energy deposited in a cone of radius $\Delta R \leqslant 0.7$ (see text) around each muon. Isolated muons are defined as those having less than 4 GeV in the cone; ▲, opposite sign, △, same sign. (*b*) A histogram of the number of muon events as a function of energy in this cone. The solid curve is the equivalent for tracks from minimum bias data and the broken curve a Monte Carlo calculation for heavy quark decays.

TABLE 1. EXPECTED ORIGINS OF EVENTS

class	number of events	probable origins
non-isolated, unlike-sign	83	heavy quark decays
non-isolated, like-sign	44	second-generation heavy quark decays plus $B^0-\overline{B^0}$ mixing
isolated, unlike-sign	57	Drell–Yan, Υ decays, and residual heavy quark decays
isolated, like-sign	15	residual second-generation heavy quark decays

with theoretical predictions (Altarelli *et al.* 1984). Figure 9 shows the data and the fitted contributions.

The non-isolated events are compared to a Monte Carlo prediction for heavy quarks (Van Eijk 1985, personal communication). Here the μ–μ jet–jet mass is plotted (figure 10) and compared with the 127 like- and unlike-sign events. Included also are the expected residual unlike- and like-sign events (21 and

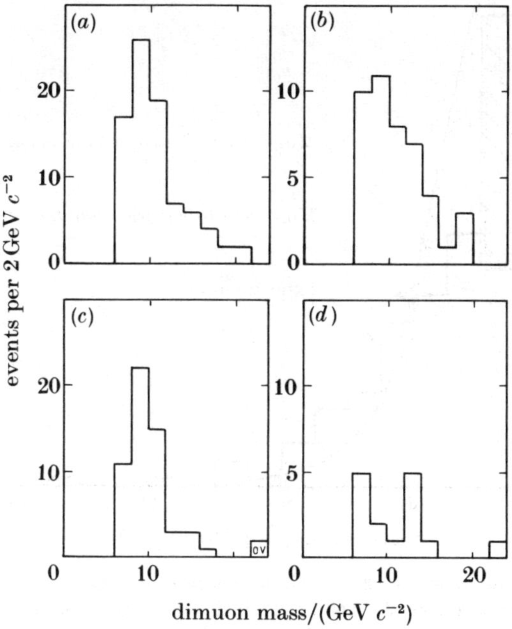

FIGURE 8. Analysis of the 199 dimuon events. (a) Non-isolated, unlike-sign, 83 events; (b) non-isolated, like-sign, 44 events; (c) isolated, unlike-sign, 57 events; (d) isolated, like-sign, 15 events.

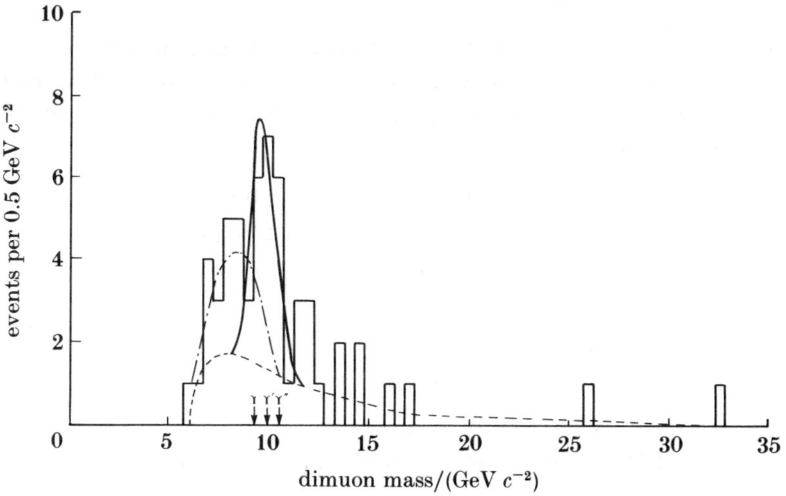

FIGURE 9. The unlike-sign, isolated dimuon sample with the estimated heavy quark contribution (chained line) and the fitted Drell–Yan (broken line) and Υ contributions (solid line).

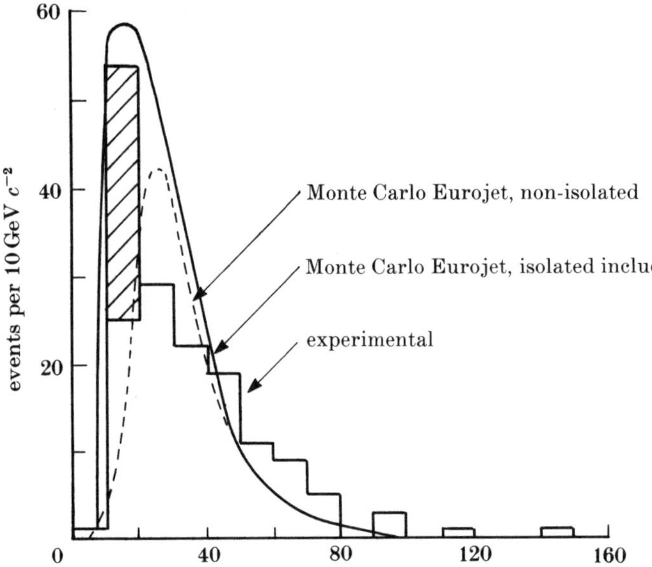

FIGURE 10. The μ–μ jet–jet mass for the 127 non-isolated events (open blocks) (unlike- and like-sign). Also shown are the 28 estimated isolated events from heavy quark decays (hatched blocks) and the corresponding Monte Carlo predictions.

7 respectively) with a corresponding Monte Carlo curve. The estimated cross section for $b\bar{b}$ production from this study, with $p_t(b) > 3$ GeV c^{-1} and $|\eta| < 2$ is

$$\sigma(p\bar{p} \rightarrow b\bar{b} + \ldots) = 1.5 \pm 1.0 \ \mu b,$$

where the error is systematic. A further indicator of heavy quark production is the strange particle content of the events as strange quarks form part of the decay chain. With certain cuts, the following figures are obtained for the number of strange particles per event.

minimum bias		0.44 ± 0.09,
$\mu^+\mu^-$	isolated	0.64 ± 0.14,
$\mu^+\mu^-$ } $\mu^\pm\mu^\pm$ }	non-isolated	1.26 ± 0.10.

We now examine the question of $B^0 - \overline{B^0}$ mixing. Defining the ratio $R = [N(++) + N(--)]/N(+-)$, we have

$R = 44/83 = 0.53 \pm 0.10$ non-isolated;
$R = 59/104 = 0.56 \pm 0.09$ all (excluding Drell–Yan and Υ).

Monte Carlo estimates with no mixing give $R = 0.33$ and with full mixing $R = 0.55$ (Ali & Jarlskog 1984; Barger & Phillips 1984). The data therefore favour mixing at the 2–3 standard deviation level.

Is there any new physics? It appears possible to account for most of what is observed within the standard model, with one exception – the like-sign, isolated events. Here, the statistics are marginal but there does appear to be an excess of events with $m_{\mu\mu} > 12$ GeV c^{-2}, with seven events observed where one would be expected (figure 11). At this stage it would be premature to say that this represented new physics, but it is clearly an interesting area to watch.

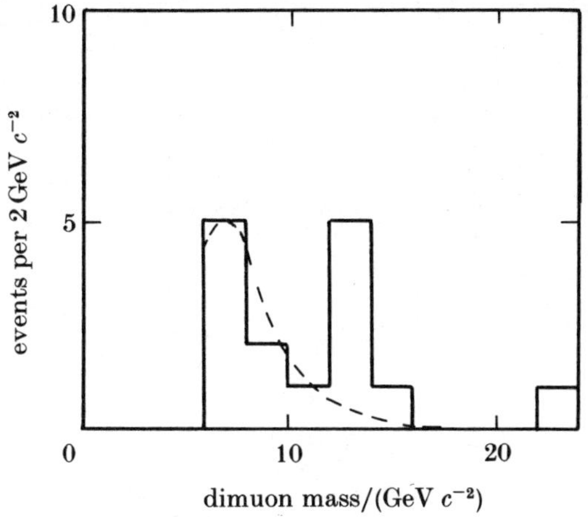

FIGURE 11. The 15 isolated like-sign dimuon events with a curve showing the expected contribution from heavy quark decays.

In summary, dimuon events have proved to be largely explicable as a combination of Drell–Yan, upsilon decay, and heavy quark production. There is some evidence that $B^0 - \overline{B^0}$ mixing is required and an indication, at a low level, of like-sign isolated events that have no ready explanation.

4. Events with missing transverse energy

Undoubtedly the most exciting observation to be made at the collider, leaving aside the W and Z, is the presence of events with 'monojets' – i.e. single jets of particles apparently unbalanced in transverse energy (Arnison *et al.* 1984*b*). The classic example of an event with missing E_T is $W \rightarrow e\nu$, where only the electron is seen and the neutrino escapes detection. This E_T imbalance has indeed proved to be a powerful indicator of W production. However, it offers the exciting prospect of searching for events containing other neutral weakly interacting particles. Now jets are not normally accompanied by neutrinos, hence the great interest in the monojet events, an example of which is shown in figure 12. One suggestion is that they might indicate the production of supersymmetric particles with the photino $\tilde{\gamma}$ (spin-$\frac{1}{2}$ partner of the photon), carrying the missing energy. The events could be the pair production of squarks $\tilde{q}$ or gluinos $\tilde{g}$ with subsequent decay yielding

$$
\begin{aligned}
p\bar{p} &\rightarrow \tilde{q}\bar{\tilde{q}} + \ldots, \\
&\rightarrow \tilde{g}\tilde{g} + \ldots, \\
&\rightarrow \tilde{g}\tilde{q} + \ldots, \\
\tilde{q} \rightarrow q\tilde{\gamma}&;\ \tilde{g} \rightarrow q\bar{q}\tilde{\gamma}.
\end{aligned}
$$

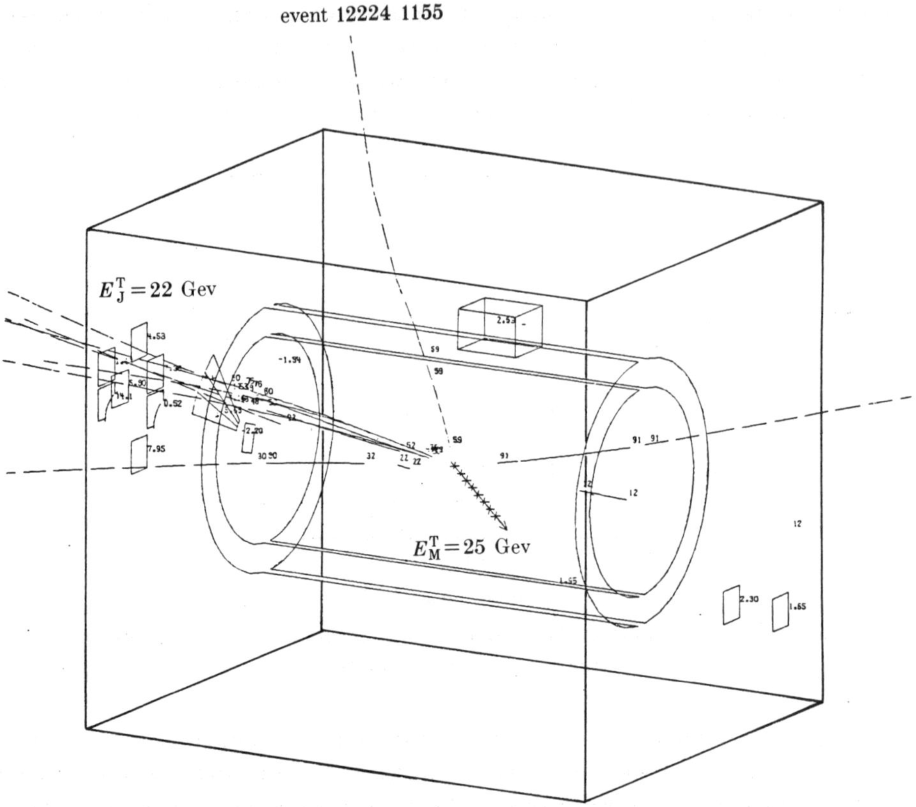

FIGURE 12. A typical 'monojet' event showing the jet and the missing E_T vector. Only tracks with $p_t > 1$ GeV c^{-1} are shown.

The squark is the spin-0 partner of the quark, and the gluino the spin-$\frac{1}{2}$ partner of the gluon. The above reactions assume that the gluino is heavier than the squark, otherwise $\tilde{q} \rightarrow q\tilde{g}$ would be the preferred decay. In general they give rise to several jets as well as missing energy. However, when experimental selection criteria are applied they can yield monojet events. Based on data taken in 1983, UA1 announced the observation of six monojet events with missing $E_T > 30$ GeV and greater than four standard deviations (Arnison *et al.* 1984*b*). One of these was classified as $W \rightarrow \tau\nu$ but the others appeared to have no standard explanation (see figure 13). On the basis of the cross section (0.06 nb) this led to a lower limit of $M_{\tilde{g}}$ or $M_{\tilde{q}} \geqslant 40$ GeV c^{-2}, assuming the events were background, following calculations by Ellis & Kowalski (1984) and others. An alternative explanation in terms of excited quarks was advanced by Kuhn & Zerwas (1984) and there have been many other theoretical papers on the topic. These will not be discussed here. Rather we shall concentrate on the existence of the phenomenon as an experimental fact and the possibility of explaining it within the framework of the standard model.

The trebling of the data in 1984 has permitted a more detailed study, especially as a 'missing energy' trigger was introduced into the hardware of the experiment,

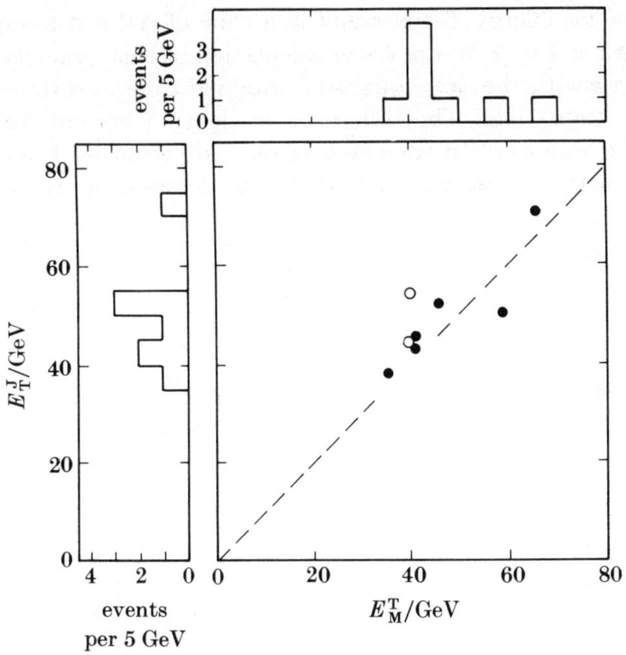

FIGURE 13. The six monojet events (●) from the 1983 data. The jet E_T is plotted against the missing E_T. Included also are two 'photon' events (○) in which one or more photons is recorded together with missing energy.

resulting in a relatively unbiased selection. The most essential experimental requirement is to be sure that the events are not the result of some malfunction or incomplete feature of the apparatus. Consequently, all candidates have been meticulously scanned to avoid situations where particles point at dead regions, etc. A particularly powerful feature of UA1 is its central detector in a magnetic field, where it can readily be established that high-energy deposits in the calorimeter have high-energy charged particles pointing towards them and that there really are no particles travelling in the opposite direction. There is no doubt that the events do exist. The 1984 data sample with E_T (jet) > 12 GeV, E_T (missing) > 4σ where $\sigma = 0.7\ (\Sigma\,|E_T|)^{\frac{1}{2}}$ (GeV) and no jet back-to-back within a 30° cone (to reduce jet–jet fluctuations) have yielded 29 monojet, 3 two-jet and 1 three-jet events, where only jets with $E_T > 12$ GeV are counted.

Having established that monojets are real it is necessary to consider all sources of such objects permitted by the standard model. The most obvious example is $W \rightarrow \tau\nu$. The τ lepton decays into hadrons and a neutrino 65 % of the time. On this basis we expect around a hundred events, before cuts, of the type

$$W \rightarrow \text{hadrons}\ \nu_\tau\,\bar{\nu}_\tau.$$

It is easy to calculate that this is in fact the main source of monojets. Experimental tools are therefore needed to distinguish $W \rightarrow \tau\nu$ from other processes. Because the τ is light (1.8 GeV c^{-2}) the jets will be narrow as the τ will have typically around 40 GeV energy – half the W mass. To characterize the narrowness we consider the fraction of transverse energy F contained in a cone of $\Delta R \leqslant 0.4$ compared with the full cone of $\Delta R \leqslant 1.0$. A Monte Carlo calculation of this quantity is shown in figure 14 compared with the same quantity obtained from experimental data on normal hadronic (QCD) jets. The difference is clear. A second handle can be obtained from the transverse momentum of the individual jet fragments. For τ decays a single particle typically carries a larger fraction of the jet transverse

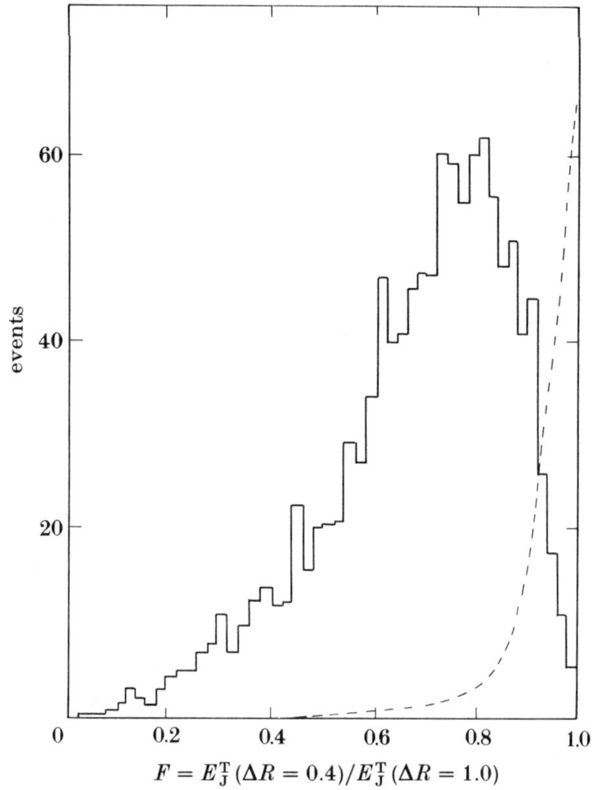

FIGURE 14. The fraction F of transverse energy deposited in a cone of $\Delta R \leqslant 0.4$ compared with that in a cone $\Delta R \leqslant 1.0$ around the jet axis. The histogram is for QCD jets (data) and the broken curve a Monte Carlo calculation for $W \rightarrow \tau\nu$ events.

momentum than in QCD jets. Figure 15 is a plot of p_t^{max}, the p_t of the highest p_t particle in a cone of $\Delta R \leqslant 0.4$, for τ jets and QCD jets. Thirdly, τ decays typically have a lower multiplicity of charged particles than QCD jets and this may also be used as a discriminator. Although the τ multiplicity is mostly less than or equal to three and nominally odd, the inclusion of a background particle can occasionally alter this. Figure 16 is a plot of the quantity F for the 29 events plotted for each multiplicity compared with a Monte Carlo expectation for $W \rightarrow \tau\nu$ events. The following cuts have therefore been chosen to define a 'τ sample':

(a) $F = \dfrac{E_T(\Delta R \leqslant 0.4)}{E_T(\Delta R \leqslant 1.0)} \geqslant 0.8;$

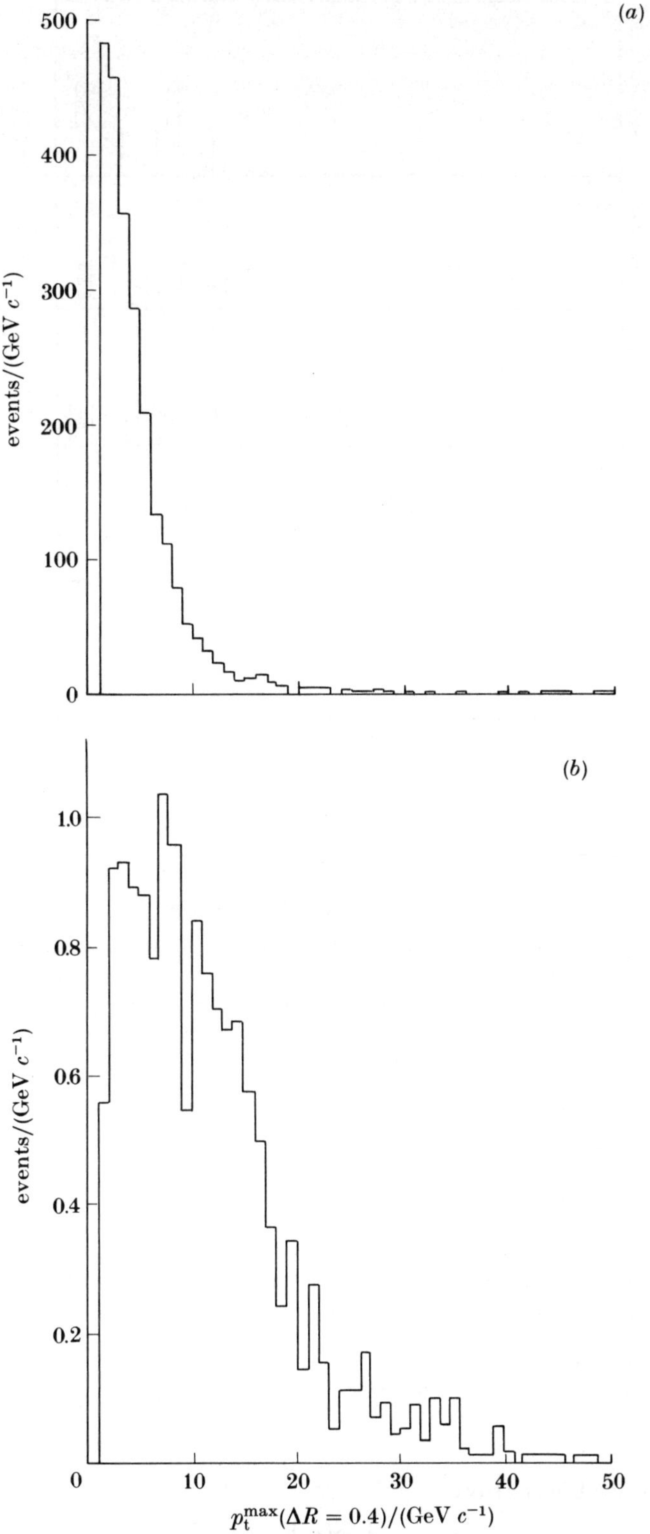

FIGURE 15. The p_t of the largest p_t particle in a cone $\Delta R \leqslant 0.4$ (a) for QCD jets and (b) for $W \rightarrow \tau\nu$ events (Monte Carlo).

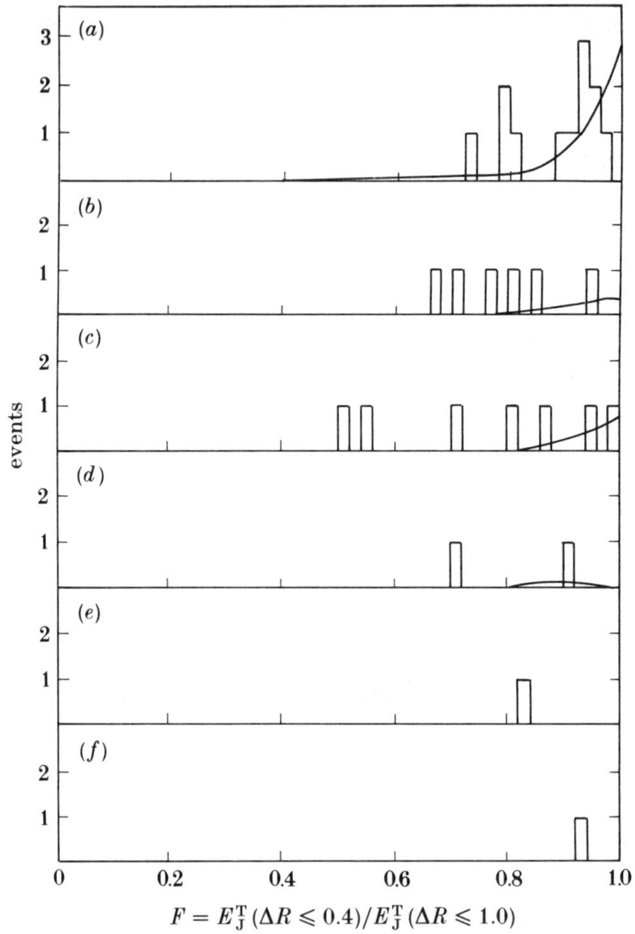

FIGURE 16. The quantity F (see figure 14) for each observed multiplicity of the 29 monojet events. The curves are the $W \rightarrow \tau\nu$ Monte Carlo expectations. (*a*) 12 one-prong events (M.C. 9.8 events); (*b*) 6 two-prong events (M.C. 1.9 events); (*c*) 7 three-prong events (M.C. 3.0 events); (*d*) 2 four-prong events (M.C. 0.5 events); (*e*) 1 five-prong event (M.C. 0.06 events); (*f*) 1 six-prong event (M.C. 0 events).

(*b*) $p_t^{max} \geqslant 3$ GeV c^{-1};

(*c*) charged multiplicity less than or equal to four.

Fourteen events pass these cuts leaving fifteen as 'monojet' candidates. It is now necessary to consider all possible standard model sources of such events. Those that have been studied at the present time are:

(i) $W \rightarrow \tau\nu$ failing the τ cuts;
(ii) $W \rightarrow e\nu + \text{jet}$, where the electron overlaps the jet;
(iii) $Z \rightarrow \nu\bar{\nu} + \text{jet}$;
(iv) $Z \rightarrow \tau\bar{\tau}$;
(v) QCD jet–jet fluctuations;

(vi) Heavy quark semi-leptonic decays in which a neutrino carries away a large amount of energy.

Present estimates from all these sources give approximately 15 events passing the τ cuts and 9 failing them. This is summarized in table 2.

TABLE 2. COMPARISON WITH STANDARD MODEL

events	passing τ cuts	failing τ cuts
standard model estimate	15	9
data	14	15

At the present level of study it is clear that a monojet signal beyond that required by the standard model cannot be excluded but does not have a high statistical significance.

Future work will include a re-analysis of the 1983 data, an extension to multijet events and a more detailed consideration of the jet topologies, improved energy calibration and improved background calculations.

The UA2 experiment is hampered by having no calorimeter coverage for angles less than 20° and so is restricted to considering jets at an E_t sufficiently high that the chance of missing a recoil jet in the apparatus is small. They have set limits for monojets as in table 3. By using UA2 cuts Ellis & Kowalski (1984) calculate $\sigma \leqslant 10$ pb for any mass of squark or gluino. Thus neither UA1 nor UA2 can exclude supersymmetry at the present time.

TABLE 3. LIMITS FOR MONOJETS (UA2)

missing E_T	cross section (90% confidence limit)
over 65 GeV c^{-1}	below 23 pb
over 50 GeV c^{-1}	below 73 pb

5. CONCLUSIONS

New physics beyond the standard model at the CERN $p\bar{p}$ collider cannot be claimed with certainty. Nor can it be excluded at the present time. When the new antiproton collector ACOL comes into operation in 1987 we can look forward to a tenfold increase in luminosity and, after two years of operation, to a 20 to 30-fold increase in statistics compared with the data described here. UA1 and UA2 will both have improved calorimeters and will be in a much better position to make decisive statements.

REFERENCES

Ali, A. & Jarlskog, C. 1984 *Physics Lett.* B **144**, 266.
Altarelli, G. *et al.* 1984 *CERN-TH.* 4015.
Arnison, G. *et al.* 1984*a* *Physics Lett.* B **147**, 493.
Arnison, G. *et al.* 1984*b* *Physics Lett.* B **139**, 115.
Arnison, G. *et al.* 1985 *Physics Lett.* B **155**, 442.

Astbury, A. *et al.* 1978 *CERN/SPSC/78-6/P92.*
Banner, M. 1985 Talk presented at '*New particles* '85', University of Wisconsin, May 8–11, 1985. (In preparation.)
Banner, M. *et al.* 1978 *CERN/SPSC/78-8/P93.*
Barger, V. & Phillips, R. J. N. 1984 *Physics Lett.* B **143**, 259.
Ellis, J. 1984 *Proceedings of the 4th topical workshop on* p$\bar{\text{p}}$ *collider physics.* Berne. CERN 84-09.
Ellis, J. & Kowalski, H. 1984 *Physics Lett.* B **139**, 115.
Kenyon, I. R. 1982 *Rep. Prog. Phys.* **45**, 1261.
Kuhn, J. H. & Zerwas, P. H. 1984 *Physics Lett.* B **147**, 189.

Discussion

P. Litchfield (*Rutherford Appleton Laboratory*). The most convincing evidence in the original UA1 claim for top was the clustering of the events around the W mass. Professor Dowell now says that half the events are produced via t$\bar{\text{t}}$ formation. Is this clustering an accident?

J. D. Dowell. Not entirely, the t$\bar{\text{t}}$ events also cluster around the W mass for kinematical reasons but less sharply than those from W-decay.

Proc. R. Soc. Lond. A **404**, 233–251 (1986)
Printed in Great Britain

The lepton–hadron interaction

By D. H. Saxon

Rutherford Appleton Laboratory, Chilton, Didcot, Oxfordshire OX11 0QX, U.K.

The scattering of leptons (e, μ, ν) by hadrons at high momentum transfer has been the incisive probe of the internal dynamics of the proton. Point-like 'partons' are found inside the protons. These have the electromagnetic and weak properties expected of quarks, and are in dynamical equilibrium with a sea of virtual quark–antiquark pairs and gluons, the quanta of the strong force.

Neutrino scattering has probed the weak coupling of the quarks, and their parity-violating handedness. Weak neutral currents (coupling with the Z^0 boson) were discovered in neutrino scattering, and parity-violating γ–Z^0 interference was observed in electron–deuteron scattering.

The higher momentum transfers that will be available at the HERA electron–proton storage ring, now under construction, will greatly expand the kinematic region available to search for new phenomena at and beyond the W, Z energy scale.

1. Introduction

In this paper we review the principal phenomena in lepton–hadron scattering, and look forward to the wide range of theoretical speculations that will be opened up to study by the construction of the HERA electron–proton colliding beam facility now underway in Hamburg.

Just as Rutherford, by scattering α-particles off gold foil, demonstrated that the atom contains a small point-like nucleus, so the phenomenon of scale invariance, or 'scaling' in deep inelastic scattering, has demonstrated the existence of point-like objects, which we will call, for the moment, 'partons', inside the proton.

The processes studied are indicated in figure 1. The point-like nature of the eeγ and μμγ couplings has been dramatically verified in the extraordinarily accurate

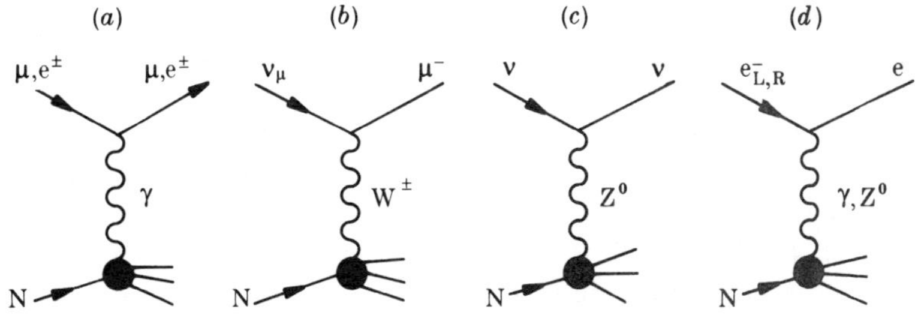

FIGURE 1. The diagrams describing deep inelastic scattering of leptons from nucleons; (*a*) electromagnetic; (*b*) weak charged-current (CC); (*c*) weak neutral-current (NC); (*d*) γ, Z° interference with an electron beam of defined helicity.

measurements of the anomalous magnetic moments of the electron and muon (Bailey *et al.* 1979; Schwinberg *et al.* 1981), in the absence of any excited state up to the highest energies available at PETRA and in searches for the residual effects of substructure in Bhabha scattering ($e^+e^- \rightarrow e^+e^-$) and $e^+e^- \rightarrow \mu^+\mu^-$, where substructure scales of the order of 1 TeV are set (Althoff *et al.* 1984; Wu 1984; Saxon 1984). ep, μp and νp scattering can therefore be used with confidence as a probe of the structure of the hadron vertex.

The weak neutral current was discovered in neutrino scattering. Figure 1*c* indicates a neutrino reaction in which there is no muon in the final state. Evidence of γ–Z^0 interference has been observed in the (parity-violating) helicity dependence of the cross section in e^-d scattering and in the (largely parity-conserving) charge asymmetry in $\mu^{\pm}C$ scattering.

2. Search for substructure in the nucleon

The story begins in Hamburg in 1933. Stern, Estermann and Frisch repeated the Stern–Gerlach experiment on hydrogen molecules, to measure the proton magnetic moment. The experiment was extremely difficult as the value expected is of the order (m_e/m_p) times that found on the electron. For the magnetic moment, $\mu = ge\hbar/2mc$, where $e\hbar/2mc$ is the Bohr magneton for mass m, and the Landé g factor, introduced to explain the anomalous Zeeman effect, would be unity if the magnetic moment were a result of orbital motion of a point charge of this mass. It was a triumph of the Dirac theory to predict $g = 2$ for the electron. Carried along by this, Pauli told Stern that if he enjoyed doing difficult experiments, he could do them, but that it was a waste of time and effort because the result was already known (Segrè 1980). The result as is now known is $g_p = 5.5856912 \pm 0.0000022$, and for the neutron, -3.8208164 ± 0.0000017 (Cohen & Taylor 1973; Greene *et al.* 1979), in complete contradiction to expectation. One concludes immediately that the proton and neutron must be composite, otherwise $g_p = 2$ is compulsory. This amazing result is readily interpreted (to about 2% accuracy) if a very naive quark model is introduced, with two quarks called $u(Q = \frac{2}{3})$ and $d(Q = \frac{1}{3})$, each point-like and of spin-$\frac{1}{2}$ (and therefore $g = 2$) and of mass 330 MeV (p = uud, n = udd). One finds $\mu_p = \frac{4}{3}\mu_u - \frac{1}{3}\mu_d$ and hence $g_p = 5.68$, and $2\mu_n = -3.77$ (see, for example, Perkins (1982)).

However, this modern interpretation was not then supported by the evidence of the symmetries of hadron spectroscopy to provide the basis for a quark model. In fact, hadron dynamics followed instead the path of Yukawa (1935) via pion exchange into Regge theory. As late as 1969, Lipkin could complain of the dichotomy between the study of dynamics (which he called s, t, u physics), where physicists vary the dynamical quantities at fixed quantum numbers, and hadron spectroscopy (I, B, Y physics), in which the quark model had a record of success, where they do the opposite (Lipkin 1969).

However, by then the first results indicating substructure in the proton had come in from ep scattering experiments. The methods and interpretation can be compared with electron–nucleus scattering at lower energies. In the 1950s, in a classic series of experiments, Hofstadter (1956) scattered electrons of energy

$E = 400$ MeV off helium and looked at the recoil-electron energy spectrum (E') as a function of angle. A clear peak from the elastic process, $e\alpha \rightarrow e\alpha$ was seen which, however, fell steeply as a function of angle, damped by a form factor that describes the probability of the α not being broken up in the scattering. At lower recoil energy there is a broad distribution, explained as scattering of the electron off a proton constituent of the α, the elastic $ep \rightarrow ep$ peak in the E' distribution being smeared by the Fermi motion of the proton within the α-particle. This does not suffer the same form factor. In modern parlance, deep inelastic $e\alpha$ scattering is elastic ep scattering.

In 4.9 GeV ep scattering at DESY, Bartel *et al.* (1968) saw a similar phenomenon; an elastic peak suppressed by a form factor, and a continuum with resonance excitation peaks at lower recoil-electron energies. If one plots the cross section, divided by the Mott formula, as a function of Q^2, then elastic scattering is seen to fall with a 'dipole' form factor,

$$\sigma_{el}/\sigma_{Mott} = G_M^2 = (1+Q^2/0.71)^{-4},$$

whereas for a final-state hadron mass of 3.5 GeV, there is almost no Q^2 dependence as Q^2 varies between 0.5 and 5 GeV (see figure 2*a*).

When presenting the SLAC results on ep scattering at 7–16 GeV, Panofsky (1968) said, 'Theoretical speculations are focused on the possibility that these data might give evidence on the behaviour of point-like, charged structures within the nucleon'.

3. Deep inelastic scattering

We start by considering elastic electromagnetic scattering of a relativistic Dirac particle ($e^{\pm}$ or $\mu^{\pm}$) from a target. By ignoring spin, one obtains the Rutherford scattering formula, $d\sigma/dQ^2 = 4\pi\alpha^2/Q^4$, where $-Q^2$ is the four-momentum transfer carried by the photon. For a relativistic electron scattering from a spin-0 target this becomes the Mott formula,

$$\left.\frac{d\sigma}{dQ^2}\right|_{Mott} = \frac{4\pi\alpha^2}{Q^4}\cos^2\tfrac{1}{2}\theta,$$

where θ is the centre-of-mass scattering angle. For a point-like fermion (Dirac) target of mass M, there is an additional term corresponding to magnetic scattering,

$$\frac{d\sigma}{dQ^2} = \left.\frac{d\sigma}{dQ^2}\right|_{Mott}\left(1+\frac{Q^2}{2M^2}\tan^2\tfrac{1}{2}\theta\right).$$

For a composite spin-$\tfrac{1}{2}$ target, such as a proton, one obtains the Rosenbluth formula,

$$\frac{d\sigma}{dQ^2} = \left.\frac{d\sigma}{dQ^2}\right|_{Mott}\left[\frac{G_E^2+(Q^2/4M^2)\,G_M^2}{1+Q^2/4M^2}+\frac{Q^2}{2M^2}\,G_M^2\tan^2\tfrac{1}{2}\theta\right],$$

where the electric and magnetic form factors, G_E and G_M, are functions of Q^2, normalized such that $G_E(0) = 1$ and $G_M(0) = 2.79 = \tfrac{1}{2}g_p$. Inspecting this formula shows that setting $G_M = 0$, a spin-0 formula is recovered, and if $G_E = G_M$ and is independent of Q^2, scattering from a Dirac-particle target is occurring.

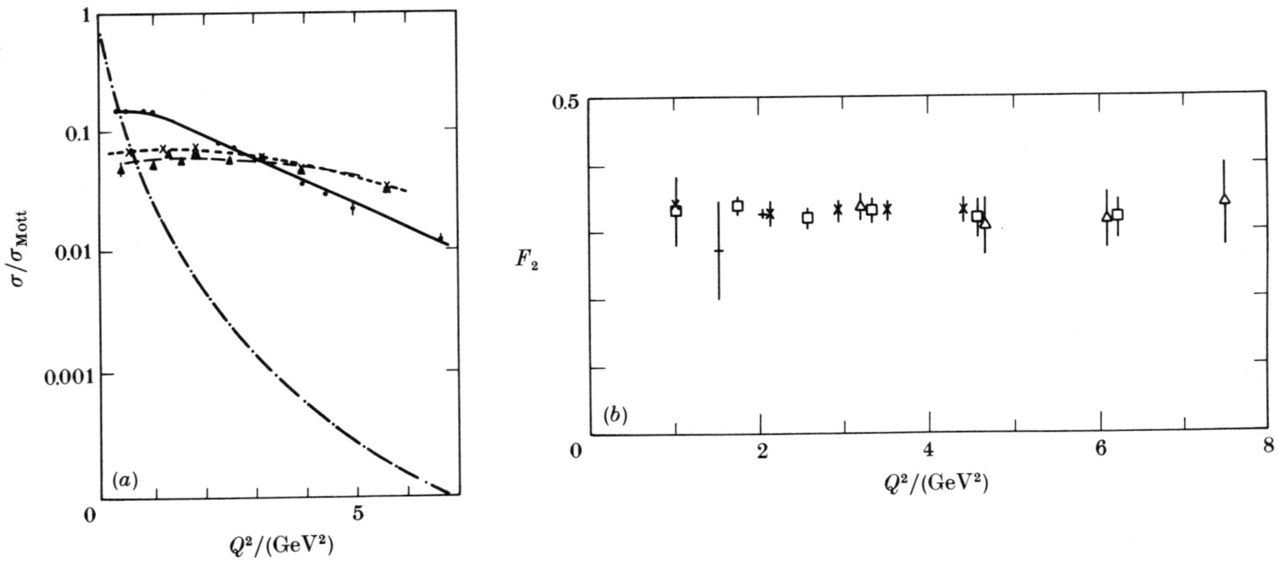

FIGURE 2. (*a*) Comparison of the cross section, divided by the Mott cross section, for elastic ep scattering (chained), and for inelastic scattering for final-state hadron masses, W = 2 (solid line), 3 (dotted), and 3.5 GeV (broken). (Llewellyn-Smith 1970; Perkins 1977.) (*b*) F_2 for the proton as a function of Q^2 at $x = 0.25$. Measurements at 6° (+), 10° (×), 18° (□), 26° (△). (Friedman & Kendall 1972.)

Consider now deep inelastic ep scattering as elastic scattering of a parton carrying a fraction x of the proton mass in its rest frame (or equivalently, a fraction x of the proton momentum in a frame in which the proton moves with high momentum). Let θ be the scattering angle in the electron–parton rest frame. Then,

$$\frac{d^2\sigma}{dQ^2\,dx} = \frac{4\pi\alpha^2}{Q^4}\frac{E'}{E}\left[\frac{F_2(x,Q^2)}{x}\cos^2\tfrac{1}{2}\theta + \frac{Q^2}{2M^2x^2}\frac{2xF_1(x,Q^2)}{x}\sin^2\tfrac{1}{2}\theta\right].$$

If the parton model is valid as a description of the proton structure it will be found for the 'structure functions', which take over the role of form factors, that $2xF_1(x,Q^2)$ and $F_2(x,Q^2)$ are functions of x only, and not of Q^2. Figure 2*b* shows $F_2(x = 0.25)$ as a function of Q^2 from SLAC results. There is no Q^2 dependence, which establishes the scale invariance of the proton structure. For spin-0 partons $F_1 = 0$ is expected, and for spin-$\frac{1}{2}$ point-like Dirac partons the Callan–Gross relation, $2xF_1 = F_2$, should hold (Callan & Gross 1969). This is indeed found to be so and it is concluded that the proton contains electrically charged spin-$\frac{1}{2}$ Dirac partons (Riordan *et al.* 1975). Can these be identified with quarks? To answer this we must look also at their weak interactions. (This discussion could have been formulated in terms of the transverse and longitudinally polarized photon cross sections σ_T and σ_L. However, to simplify it I have anticipated the result on the Callan–Gross relation, because for highly relativistic spin-$\frac{1}{2}$ partons, $\sigma_L = 0$, and suppressed this aspect (see Friedman & Kendall 1972). The ratio $R = \sigma_L/\sigma_T$ is still poorly known, but is below about 0.2 for $x > 0.4$ (Dydak 1983). The EMC for μp scattering have measured $\langle R\rangle = -0.010 \pm 0.037 \pm 0.102$ (systematic), consistent with zero, or with the small value predicted by quantum chromodynamics, QCD (Aubert *et al.* 1985*a*; Fisk & Sciulli 1982; Bardeen *et al.* 1978).)

The total cross sections for the reactions $\nu_\mu N \rightarrow \mu^- + \text{hadrons}$, and $\bar{\nu}_\mu N \rightarrow \mu^+ + \text{hadrons}$, have been measured at incident moments of 280 and 200 GeV c^{-1} respectively. Although many different hadronic final states are seen, the total cross sections are both linearly proportional to the neutrino energy (Wotschak 1981). This is expected of a four-fermion point interaction. The lack of dimension in the energy dependence suggests that the proton, as seen by a weak-interaction probe, contains point-like, spin-$\frac{1}{2}$ partons.

To establish the connection with electromagnetic scattering, we recapitulate the formalism. Let $-Q^2$ represent four-momentum transfer from the initial lepton and $x = Q^2/2M(E-E')$, for a target of mass M at rest where E, E' are the initial- and final-state lepton energies. Then x is the fraction of the mass carried by the parton. Let $y = \sin^2\frac{1}{2}\theta$ in the lepton–parton centre-of-mass frame. Then $Q^2 = 2MExy$, and $e^{\pm}N$ and $\mu^{\pm}N$ scattering can be rewritten as,

$$\frac{d^2\sigma}{dQ^2\,dx} = \frac{4\pi\alpha^2}{Q^4}\left[(1-y)\frac{F_2^{eN}}{x} + \tfrac{1}{2}y^2\frac{2xF_1^{eN}}{x}\right].$$

With a (V − A) coupling, νQ and $\bar{\nu}\bar{Q}$ scattering (Q = parton) are isotropic, and $\bar{\nu}$Q and ν$\bar{Q}$ scattering have a $(1-y)^2$ distribution. For a target containing equal numbers of neutrons and protons, as commonly used in experiments (D_2, $CaCO_3$,

SiO_2), we take advantage of charge symmetry. Defining $F_i^{\nu N} = \frac{1}{2}(F_i^{\nu n} + F_i^{\nu p})$, then $F_i^{\nu N} = F_i^{\bar{\nu} N}$, and

$$\frac{d^2\sigma}{dQ^2\, dx} = \frac{G^2}{2\pi}\left[(1-y)\frac{F_2^{\nu N}}{x} + \tfrac{1}{2}y^2 \frac{2xF_1^{\nu N}}{x} \pm y(1-\tfrac{1}{2}y)\frac{xF_3^{\nu N}}{x}\right].$$

A third structure function is required because of the parity violation in the weak interaction. The positive sign holds in νN scattering and the negative sign in $\bar{\nu}$N scattering. Examination of the angular (y) dependence yields $F_2^{\nu N}/x = 2[Q(x) + \bar{Q}(x)]$ and $F_3^{\nu N} = 2[Q(x) + \bar{Q}(x)]$. Further, if partons are quarks, and the proton may contain u, d, ū and d̄ quarks (ignoring s, s̄, c and c̄ quarks),

$$F_2^{\nu p}(x) = 2x[d(x) + \bar{u}(x)],$$

therefore $$F_2^{\nu N}(x) = x[u(x) + \bar{u}(x) + d(x) - \bar{d}(x)],$$

where $u(x)$ is the x distribution of u quarks in the proton, and therefore, by charge symmetry, of d quarks in the neutron. For ep scattering, one obtains, including factors for quark charge-squared,

$$F_2^{ep}(x) = x\{\tfrac{4}{9}[u(x) + \tilde{u}(x)] + \tfrac{1}{9}[d(x) + \bar{d}(x)]\}.$$

Hence $$F_2^{eN}(x) = \tfrac{5}{18}x(u + \tilde{u} + d + \bar{d}),$$

and finally, $$F_2^{\nu N}(x, Q^2) = \tfrac{18}{5}F_2^{eN}(x, Q^2).$$

The results are shown in figures 3 and 4. In figure 3 we show a composite of results from several μN and νN experiments. Normalization factors of the order of 10 % are

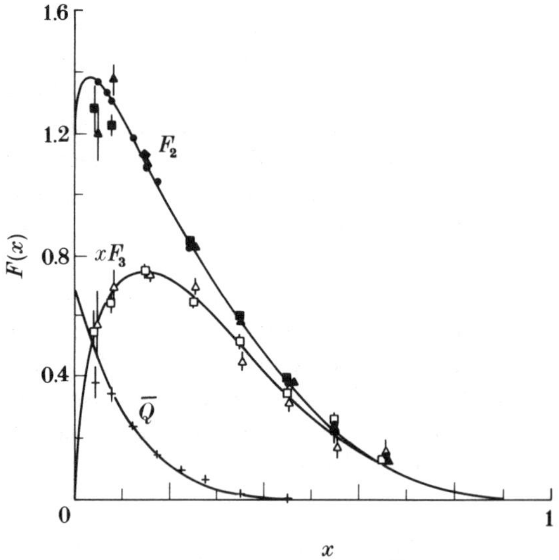

FIGURE 3. Structure functions F_2, xF_3 and $\bar{Q}$ measured in different experiments, for $10 < Q^2 < 30$ GeV², assuming $R = \sigma_L/\sigma_T = 0$. • Electromagnetic measurements: EMC (Aubert *et al.* 1981) multiplied by 18/5: ◆, BFP (Clark *et al.* 1983; Meyers 1983) multiplied by 18/5 × 0.95. Neutrino measurements, ■□+, CDHS (Abramowicz *et al.* 1983) multiplied by 1.07; ▲△, CCFRR (Particle Data Group 1984).

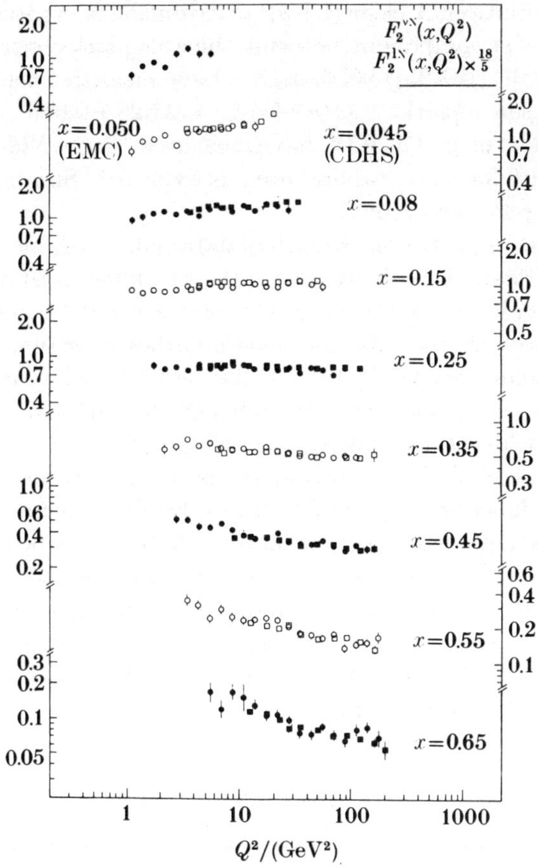

FIGURE 4. The structure function F_2, as a function of x and Q^2 for neutrino data (CDHS experiment, ●○) and μN scattering, multiplied by 18/5 (EMC experiment, ■□). (Dydak 1983.)

needed between them. The momentum weighted distributions of $(Q+\overline{Q})$, $(Q-\overline{Q})$, and hence $\overline{Q}$ are shown. The antipartons in the nucleus are seen to be concentrated at low x. From these results the total number of valence partons can be measured (Gross & Llewellyn-Smith 1969). Different experiments find

$$N(\mathrm{Q})-N(\overline{\mathrm{Q}}) = \int F_3\,\mathrm{d}x = 3.2\pm0.5,\ 3.13\pm0.48,\ 2.33\pm0.15\pm0.09\pm0.10,$$

consistent with a naive quark–parton model expectation of 3 (lower by about 7% in QCD (de Groot *et al.* 1979; Fisk & Sciulli 1982; Bolognese *et al.* 1983; Bodek 1984)). The total fractional mass or momentum carried by the charged partons, $\int F_2\,\mathrm{d}x$, is about 0.5. This initially surprising result shows that half the momentum of the proton is carried by non-interacting neutral particles.

By comparing ν and $\bar{\nu}$ scattering on H_2 and other elements (Dydak 1983), it is possible to extract separately the valence quark distribution $u_v = u-\bar{u}$ and d_v, and the 'sea' contributions $(\bar{u}+\frac{3}{4}s)$, $(\bar{d}+\frac{3}{4}s)$. One takes advantage of the difference in the angular distributions of νQ and ν$\overline{\mathrm{Q}}$ scattering. For example, in $\bar{\nu}$N scattering the reactions at the quark level are $\bar{\nu}\mathrm{d}\rightarrow\mu^+\bar{\mathrm{u}}$, with a flat distribution, and $\bar{\nu}\mathrm{u}\rightarrow\mu^+\mathrm{d}$, with a $(1-y)^2$ distribution. A clear $(1-y)^2$ distribution is evident in the data, but with an underlying flat component, showing the antiquark contribution (de Groot *et al.* 1979). The BEBC (WA25) Collaboration have measured the total momentum carried by the sea quarks, $\int x(\bar{u}+\frac{3}{4}s)\,dx = 0.028 \pm 0.006$ and $\int x(\bar{d}+\frac{3}{4}s) = 0.028 \pm 0.008$ (Dydak 1983). The EMC have measured $N(u)-N(d) = 3\int(F_2^{\mu p}-F_2^{\mu n})/x\,dx = 0.72 \pm 0.06 \pm 0.39$ (syst), where unity is expected. Sum rules are discussed at greater length by Perkins (1977).

Figure 4 represents a part of the wealth of data now available. There are several features to notice. First, the νN and μN points, the latter multiplied by 18/5, are indistinguishable at all x and all Q^2, in spectacular verification of the quark–parton model. Secondly, around $x = 0.25$, the scale invariance, shown in earlier work in figure 2, is verified and extended up to $Q^2 \approx 200\ \text{GeV}^2$. Thirdly, as Q^2 is varied there is, in general, a slow scaling violation. At higher Q^2 the parton distribution becomes softer, increasing at low Q^2 and decreasing at high Q^2.

The evidence presented above emboldens us to identify partons with quarks, based on scale invariance ($F_i(x, Q^2) = F_i(x \text{ only})$), the Dirac couplings of the partons ($2xF_1 = F_2$), the relative electromagnetic and weak couplings being consistent with 5/18 and independent of x and Q^2, and the number of valence partons being consistent with three. We see the parton as containing three valence quarks (u, u and d) and a sea of soft q$\bar{\text{q}}$ pairs. Two puzzles remain. Only half of the proton momentum is accounted for, and there is a slow evolution with Q^2 of the structure function. This forms the topic of §4.

4. The role of gluons

The missing component of the proton that has neither electromagnetic nor weak interactions, has appeared so far only in a passive role. We now look at the gluons in a more active way. The gluons act in QCD as the vector-field quanta analogous to the photons in electromagnetism. These are responsible for binding the quarks and antiquarks together to form the proton. We concentrate here on their role in the evolution of the structure function.

When an electric charge is accelerated, for example by passage through matter, it emits bremsstrahlung photons, and photons striking matter can create e^+e^- pairs. In exactly the same way, when a proton is impacted by the Q^2 of the deep inelastic scatter, the following processes can take place: q→qg, g→q$\bar{\text{q}}$ and the self-coupling of the gluons, g→gg. In each of these splittings the number of quanta increases and the momentum is shared, so that one expects the structure function to become softer and to contain more sea q$\bar{\text{q}}$ pairs as Q^2 is increased, and less of the momentum to be carried by valence quarks. There is at any Q^2 an equilibrium between the x distribution of the quarks, antiquarks, and gluons, and the change in this equilibrium with a small change in Q^2 is described by the Altarelli–Parisi equations (Altarelli & Parisi 1977):

$$Q^2 \frac{\partial}{\partial Q^2} \begin{pmatrix} q(x, Q^2) \\ \bar{q}(x, Q^2) \\ G(x, Q^2) \end{pmatrix} = \frac{\alpha_s(Q^2)}{2\pi} \int_x^1 \frac{dy}{y} P(y) \begin{pmatrix} q(x/y, Q^2) \\ \bar{q}(x/y, Q^2) \\ G(x/y, Q^2) \end{pmatrix},$$

where the 3×3 matrix P describes the splitting function for an elementary process such as $g \rightarrow q\bar{q}$. Thus, in analogy with an electromagnetic shower,

$$P_{qG}(z) = \tfrac{1}{2}[z^2 + (1-z)^2],$$

$$P_{Gq}(z) = \tfrac{4}{3}[1 + (1-z)^2]/z,$$

$$P_{qq}(z) = \tfrac{4}{3}[(1+z^2)/(1+z) + \tfrac{3}{2}\delta(1-z)],$$

and

$$P_{gG}(z) = 6[z/(1-z) + (1-z)/z + z(1-z) + (\tfrac{11}{12} - \tfrac{1}{8}f)\,\delta(1-z)],$$

where f is the number of active flavours. (The handling of the singularities is a little subtle; see, for example, Buras (1980); Dokshitser *et al.* (1978).)

Experimentally, these integro–differential equations must be solved to fit the measured structure functions as a function of Q^2. The scale of the Q^2 variation is given by the strong-coupling constant, where, to leading order,

$$\alpha_s = \frac{12\pi}{(33-2f)\ln(Q^2/\Lambda^2)}$$

Modern analyses try to solve the equations directly (see, for example, Devoto *et al.* (1983); Duke & Roberts (1984)). Acceptable fits are obtained to a wide range of data in νN, μN and μp scattering, and the Q^2 evolution of F_2 and xF_3 is described well. The ability to extract the unknown Λ and the gluon structure function $G(x, Q_0^2)$ at fixed Q_0 is more limited. The uncertainty in Λ is understandable as for most data $Q^2 \gg \Lambda^2$. Also, as Q^2 approaches Λ^2, α_s blows up and the theory, usually carried to second order in α_s (so that terms such as $P_{q\bar{q}}(z)$ are needed), becomes inapplicable. One finds solutions for Λ in the range 100–300 MeV, giving $\alpha_s \approx 0.2$. The gluon distribution at low Q^2 is not well determined. Duke & Owens (1984) find acceptable fits with $xG(x, Q_0^2 = 4) = 1.564\,(1+9x)(1-x)^6$, or $0.879(1+9x)(1-x)^4$. When evolved to the region, $Q^2 = 90$, in the midst of the data set, these are virtually indistinguishable.

5. Partons in nuclei – the EMC effect

It came as a great surprise when the European Muon Collaboration (EMC) published their measurement of the ratio $F_2(x, \mathrm{Fe})/F_2(x, \mathrm{D}_2)$ in μN scattering at 280 GeV, as shown in figure 5 (Aubert *et al.* 1983). In very little time, however, it was verified by some physics ‘archaeology’ on old Rochester–SLAC–MIT data (Bodek *et al.* 1983), looking at the target-empty data, and has subsequently been verified over a range of target materials from He to Au (Arnold *et al.* 1983; Dydak 1983; Savin 1984). This behaviour is in complete contrast to the expectations of Fermi smearing. The same phenomenon has been searched for in neutrino scattering but the results are inconclusive (Savin 1984; Abramowicz *et al.* 1984; Sarkar-Cooper 1984). At large x the effect is confirmed with modest significance, but at low x none of the experiments see any effect. Recent BCDMS data on μN are also consistent with no enhancement at low x (Feltesse 1985).

Although the effect was not anticipated, a wealth of interpretations are now

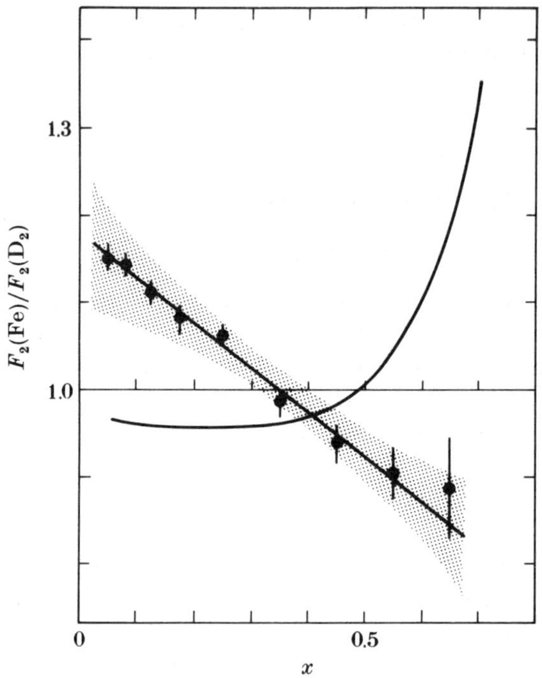

FIGURE 5. The ratio of the nucleon structure function $F_2^{\mu N}$ measured on Fe and on D_2; EMC data. Shaded band indicates systematic error, common to all points. Rising curve shows the effect of Fermi motion alone.

available. Savin (1984) classified these into two types – a change of nucleon parameters, or multiquark states in nuclei. I mention here just two ideas. Close *et al.* (1983, 1984) suggested that the confinement radius of the proton is larger in a nucleus by about 15%. If you like, the bag surface tension is reduced when immersed in nuclear matter. Llewellyn-Smith (1983) and Ericson & Thomas (1983) suggested that nuclei contain about 0.12 A pions.

Experimental indications in support of the effect have come from another measurement by the EMC. They have studied the process $\mu N \rightarrow \mu\psi + \ldots$. The ψ is created by photon–gluon fusion (Gluck & Reya 1978, 1979; Leveille & Weiler 1979). A gluon in the proton materializes under the Q^2 of the photon as a $c\bar{c}$ pair that forms the ψ. They found that at very low $x = M^2_{\psi}/2M(E-E') \approx 0.06$, $G(\mathrm{Fe})/G(\mathrm{H, D}) = 1.45 \pm 0.12 \pm 0.22$ (Aubert *et al.* 1985*b*). Photon–gluon fusion (excluded from this paper for reasons of space), is thought to be the main source of charmed particles in μN scattering. An intrinsic $c\bar{c}$ contribution to the nucleon sea was predicted by Brodsky *et al.* (1980), but this predicts a substantially harder x distribution for dimuon production (incident $\mu + \mu$ from charm decay) than is observed (Drees 1981; Aubert *et al.* 1981*b*).

6. Weak neutral currents

We turn now to the other area where lepton–nucleon scattering has greatly advanced our knowledge, in the discovery and phenomenology of weak neutral currents, via the exchange of the Z° boson (figure 1*c*, *d*). Weak neutral currents are known to be strongly suppressed in $\Delta s \neq 0$ decays. The upper limits on the branching ratios of the decays $K^\pm \to \pi^\pm e^+e^-$, $\pi^\pm\mu^+\mu^-$ and $\pi^\pm\nu\bar{\nu}$ are of the order of 10^{-7} to 10^{-6}, for $K^0_L \to e^+e^-$ and $e^\pm\mu^\mp$, 2×10^{-7} and 6×10^{-6} respectively, and the rate for $K^0_L \to \mu^+\mu^-$ is $(9.1\pm0.9)\times10^{-9}$. In 1972, measurements of $(\nu p \to \nu p)/(\nu n \to \mu^- p)$ and $(\nu p \to \nu n\pi^+)/(\nu p \to \mu^- p\pi^+)$ suggested only low rates (Perkins 1972). In contrast to this, the theoretical interest for neutral currents was very high. Glashow (1961), Weinberg (1967) and Salam (1968) had constructed a unified theory of weak interactions which was shown by 't Hooft (1971*a*, *b*) to be renormalizable. In this theory, as well as W^+, W^- and γ, a Z° boson is necessary. The Z° coupling to the neutrino is pure V–A, but the vector couplings to charged leptons and quarks are mixed with photon couplings. The mixing is described by a single parameter, the so-called Weinberg angle, θ_W. The fermion Z° couplings are then as given in table 1, and the W and Z masses are given by

$$m_W = \left(\frac{\pi\alpha}{\sqrt{2}\,G_F(1-\Delta r)}\right)^{\frac{1}{2}} \frac{1}{\sin\theta_W}, \quad m_Z = m_W/\cos\theta_W,$$

where G_F is determined from μ decay. The quantity Δr contains one-loop corrections of the order of 6% (Marciano 1984). Notice that for $\sin^2\theta_W$ close to 0.25, as is found experimentally, the lepton$^\pm$ Z° coupling is almost purely axial.

Table 1. Vector and axial weak neutral couplings[a]

	Q	g_a	g_v	
ν_e, ν_μ, ν_τ	0	$\frac{1}{2}$	$\frac{1}{2}$	$= 0.5$
e, μ, τ	-1	$-\frac{1}{2}$	$-\frac{1}{2}+2\sin^2\theta_W$	$= -0.06$
u, c, t	$\frac{2}{3}$	$\frac{1}{2}$	$\frac{1}{2}-\frac{4}{3}\sin^2\theta_W$	$= 0.21$
d, s, b	$-\frac{1}{3}$	$-\frac{1}{2}$	$-\frac{1}{2}+\frac{2}{3}\sin^2\theta_W$	$= -0.35$

[a] The numerical values given for $\sin^2\theta_W = 0.22$. Left- and right-handed fermion couplings are related to the above by $g_L = \frac{1}{2}(g_v+g_a)$, $g_R = \frac{1}{2}(g_v-g_a)$.

Experimentally, the first results came in neutrino scattering. A single $\nu_\mu e^- \to \nu_\mu e^-$ scatter was observed by the Gargamelle Collaboration (Hasert *et al.* 1973*a*), and results on $\nu N \to \nu X$ followed (Hasert *et al.* 1973*b*; Benvenuti *et al.* 1974; Aubert *et al.* 1974; Barish, S. J. *et al.* 1974; Barish, B. C. *et al.* 1975). Experimentally this is quite difficult as one has to be convinced of two things. First that an outgoing muon has not escaped detection in a reaction of the type $\nu_\mu N \to \mu^-, \ldots$, and secondly that the incident particle was a neutrino and not, say, a neutron or K^0_L, perhaps produced in an upstream neutrino reaction. With the construction of large detectors, these 'edge effects' have become easier to control, and several experiments have now measured the ratios $R_\nu = \sigma_\nu(NC)/\sigma_\nu(CC)$ and $R_{\bar{\nu}} = \sigma_{\bar{\nu}}(NC)/\sigma_{\bar{\nu}}(CC)$, where NC and CC stand for neutral- and charged-current reactions. The results are summarized in figure 6. The theoretical expressions are $R_\nu = (u_L^2+d_L^2)+\frac{1}{3}(u_R^2+d_R^2)$, $R_{\bar\nu} = (u_L^2+d_L^2)+3(u_R^2+d_R^2)$. The evaluation is quite complicated. Charged- and neutral-current interactions on sea quarks have to be included and the effects of quark masses in flavour-charging transitions included. Other ratios, such as $R' = [\sigma_\nu(\mathrm{NC})-\sigma_{\bar\nu}(\mathrm{NC})]/[\sigma_\nu(\mathrm{CC})-\sigma_{\bar\nu}(\mathrm{CC})] = \frac{1}{2}-\sin^2\theta$, have lower systematic errors as the strange quark sea effects largely cancel. All recent data may be summarized by $\sin^2\theta_W = 0.225\pm0.007$ (exp) ±0.006 (theory) (Geweniger 1984).

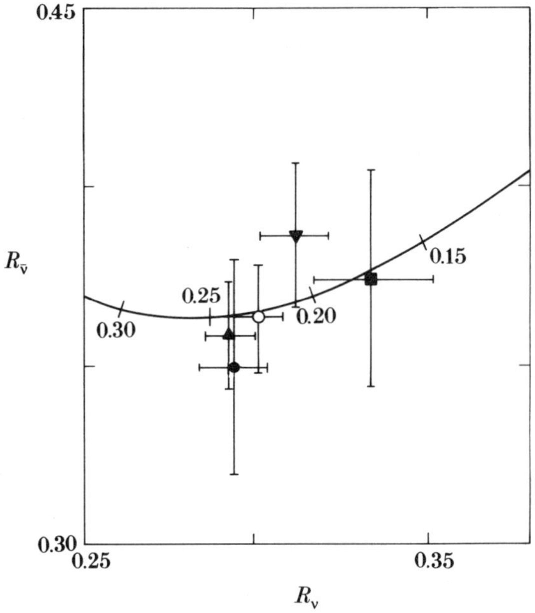

FIGURE 6. Measurements of $R_\nu = \sigma_\nu(\mathrm{NC})/\sigma_\nu(\mathrm{CC})$ and $R_{\bar\nu}$ in νN reactions from five experiments: ●, CDHS 1977; ▼, CHARM 1981; ■, BEBC 1983; ▲, CCFRR 1984; ○, CDHS 1984; all scaled to the conditions of the CDHS experiment. Curve shows the locus expected as $\sin^2\theta_W$ varies (Geweniger 1984).

Because of the uncertainties arising from hadron structure, a new experiment is underway to measure $R_e = \sigma(\nu_\mu e)/\sigma(\bar\nu_\mu e)$:

$$R_e = 3\,\frac{1-4\sin^2\theta+\frac{16}{3}\sin^4\theta}{1-4\sin^2\theta+16\sin^4\theta}.$$

Although the statistic is harder to come by (cross section $\approx G^2 m_e E_\nu$ and not $G^2 m_p E_\nu$), there are no hadronic corrections and an error of ±0.005 is anticipated on $\sin^2\theta_W$ (Winter 1984).

Following the successful observation of weak neutral currents in neutrino interactions, it was natural to look for an effect in charged lepton interactions. Parity violation has been observed in forbidden M1 tansitions in atoms (Commins & Bucksbau 1980; Bouchiat *et al.* 1984) at $Q^2 = 10^{-18}\ \mathrm{GeV}^2$, i.e. *ca.* 1 eV^2. Electroweak interference asymmetries are all proportional to the ratio of the weak and electromagnetic propagators, and hence, at low Q^2, to Q^2/M_Z^2. They are enhanced here by careful selection of forbidden transitions, which reduce the relative magnitude of the electromagnetic coupling.

A dramatic demonstration came in 1978 with the dependence of the cross section for deep inelastic e^-d scattering on incident electron helicity. Any such dependence is parity-violating and demonstrates the interference of γ exchange with Z° exchange. Careful experiments were made at beam energies of 16–22 GeV c^{-1}, taking advantage of the change of the precession of the electron spin about the beam transport system (owing to the anomalous magnetic moment). A rotation of 2π is observed between 16.2 and 22.2 GeV c^{-1}. The electrons were produced with 37 % polarization by back-scattering them from a circularly polarized laser beam. The result is $A = (\sigma_R - \sigma_L)/(\sigma_R + \sigma_L) = -(9.5 \pm 1.6)10^{-5}\, Q^2$, for $Q^2 \approx 5$ GeV^2. This established the presence of a weak axial current coupling to the electron, and from this asymmetry and from its y dependence, we can measure $\sin^2\theta_W = 0.224 \pm 0.020$ (Prescott *et al.* 1978, 1979).

A charge asymmetry in the y distribution was observed in 1982 by the BCDMS (Argento *et al.* 1983) in $\mu^\pm C \rightarrow \mu^\pm X$, at 120 and 200 GeV c^{-1}. The charge asymmetry is a function of y and Q^2, as follows:

$$B(y, Q^2) = (\sigma^+ - \sigma^-)/(\sigma^+ + \sigma^-)$$
$$= -1.8\ 10^{-4} \Gamma_\mu \tfrac{6}{5}(a_d - 2a_u)\, g(y) Q^2,$$

where
$$\Gamma_\mu = (a_\mu - |P_\mu| v_\mu)$$

and
$$g(y) = [1-(1-y)^2]/[1+(1-y)^2].$$

(The beam helicity $|P_\mu| = 0.66$ or 0.81.) Because $v_\mu (= g_V(\mu))$ is close to zero, this is almost a parity-conserving asymmetry resulting from the axial couplings of the lepton and the quarks. One per cent asymmetries were seen, in agreement with theory. From the magnitude of the asymmetry one obtains a value for v_μ and hence $\sin^2\theta_W = 0.23 \pm 0.07 \pm 0.04$. Interference of one- and two-photon exchange diagrams also generates a charge asymmetry. However, this falls with Q^2 and so is not confused with the electroweak interference asymmetry.

The Weinberg angle has been measured now in a variety of processes over a wide range of Q^2 as given in table 2. A single value fits all processes and with the discovery of the W and Z at the expected masses, constitutes an outstanding confirmation of the theory.

TABLE 2. MEASUREMENTS OF THE WEINBERG ANGLE[a]

process	$Q^2/(\text{GeV}^2)$	$\sin^2\theta_W$
parity violation in Cs	10^{-18}	$0.205 \pm 0.034 \pm 0.024$
$\nu_\mu e/\bar{\nu}_\mu e$	10^{-2}	$0.215 \pm 0.032 \pm 0.015$
$e^-_{L,R}d$	10^0	0.224 ± 0.020
$\nu_\mu N$	10^1	$0.225 \pm 0.007 \pm 0.006$
$\mu^\pm N$	10^2	$0.23 \pm 0.07 \pm 0.04$
$e^+e^- \rightarrow \mu^+\mu^-, \tau^+\tau^-$	10^3	0.18 ± 0.03
M_W, M_Z	10^4	0.221 ± 0.007

[a] Data from Bouchiat *et al.* (1984); Saxon (1984); Marciano (1984); Bergsma *et al.* (1984); Prescott *et al.* (1978), (1979); Argento *et al.* (1983); Geweniger (1984).

7. Future prospects

Our present understanding of elementary particles as composed of quarks and leptons interacting through spin-1 bosons rests on two historic foundations – the beautiful range of hadronic spectroscopy, and the dynamics of lepton–nucleon collisions. Since this, now conventional, wisdom was set up, we have seen the beautiful appearance of jets, and in particular of gluon jets in e^+e^- collisions (see, for example, Söding & Wolf 1981), and the discovery of the gauge bosons $W^{\pm}$ and Z°. What then will lepton–nucleon collisions contribute in the future?

A new electron–proton colliding beam facility is currently under construction in Hamburg. Hera, as it is called (hadron electron ring accelerator) is an underground circular tunnel of 6 km circumference containing counter-rotating electron and proton beams in separate magnet rings, arranged to intersect at four points around the ring, where experiments can be constructed. Beams of electrons or positrons, with either helicity at up to 30 GeV, collide head-on with protons of 820 GeV. This gives a maximum Q^2 of 98000 GeV^2, equivalent to an electron beam of 51 TeV incident on a stationary target. The kinematic range accessible is thus increased by more than an order of magnitude over present facilities. The luminosities expected are such that $e_L p \rightarrow \nu X$ will occur with 100 events per day. Several countries are contributing to the design and construction and first collisions are expected in 1990. For recent descriptions of the facility and possible experiments see HERA reports (1983, 1985). Hera will extend our knowledge of quark and boson physics, but more particularly, will open up new physics possibilities on the Fermi mass scale.

The structure functions will continue to shrink to lower x. Gluon bremsstrahlung should give rise to clear multijet events, allowing the measurement of α_s at $Q^2 > 10000$. We should measure $\alpha_s \approx 0.08$, well below present values, which will provide direct evidence that the strong coupling constant really does 'run' with Q^2. The γ–Z° interference will clearly influence the y distributions. For $y = 0.8$ the ratio $\sigma(\gamma + Z^{\circ})/\sigma(\gamma)$ at $x = 0.25$ will range from 2.5 (for e_L^- incident) to 0.3 for left-handed positrons (Perkins 1983). There is ample room to check out detailed predictions of the standard model, or to look for variations around it.

The main QCD neutral current processes are sketched in figure 7. They are, first, deep inelastic scattering, which gives rise to one high p_T jet with a leading u or d quark, and secondly, QCD Compton scattering, with an additional gluon jet; thirdly, photon–gluon fusion, which is a source of heavy flavours not present

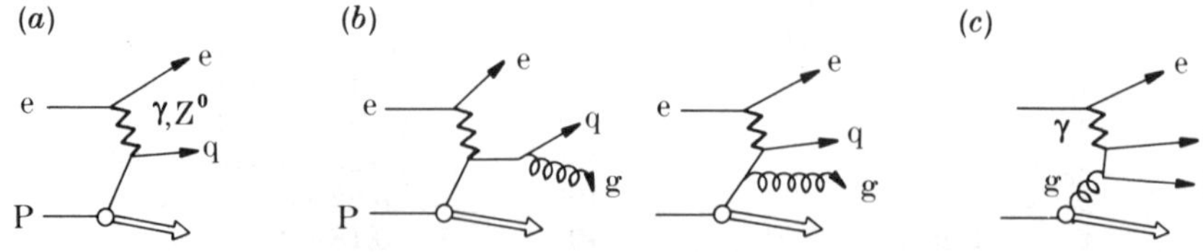

Figure 7. The main neutral current QCD processes. (a) Deep inelastic scattering; (b) Compton scattering (Gluon bremsstrahlung); (c) photon–gluon fusion. This is the main source of heavy flavours ($c\bar{c}$, $b\bar{b}$) not present as constituents of the proton.

in the target proton. At $Q^2 \approx 500\ \mathrm{GeV}^2$ these should occur in the ratio 2:2:1 (approximately). Almost 40% of the γ–g fusion events should be $c\bar{c}$ events and 5% $b\bar{b}$, giving a copious $b\bar{b}$ yield (Longo 1983; R. Thompson 1984, personal communication).

The standard model of quarks, leptons and bosons has had widespread success, and prompted many theoretical attempts to explain or extend it. We look briefly now at the impact of HERA on these speculations, for new physics at or beyond the Fermi mass scale (Cashmore 1983; Cashmore *et al.* 1985; Bagger & Peskin 1985). Leptoquarks (quark–lepton bound states) are expected in theories in which the quarks and leptons have a common origin (such as extended technicolour (Lane & Peskin 1980; Fahri & Susskind 1981)), or the W and Z bosons are composite (Schrempp & Schrempp 1985). Chadha *et al.* (1983) have studied the production and identification of the predicted pseudoscalar lepton quarks of mass 160 GeV. Because the technicolour force takes over the role of the Higgs particle the leptoquark couplings are proportional to the masses, so that the dominant production is by electron–gluon fusion, in which the gluon makes a virtual $t\bar{t}$ pair and $e\bar{t}$ form a P_{LQ} resonance, giving a $\tau t\bar{t}$ final state. The reconstructed lepton-jet invariant-mass distribution shows a clear peak at the P_{LQ} mass (even allowing for missing neutrinos), well resolved from the background because of γg fusion. Signatures such as $\mu t\bar{t}$ (μ from τ decay) are in themselves very striking. Depending on the mixing angles, masses of 180 GeV should be accessible. In the scheme of Schrempp & Schrempp, pseudoscalar leptoquarks exist at 20 GeV, and may already have been seen, and vector leptoquarks would be copiously produced, even at a mass of 250 GeV. Similar invariant-mass techniques can be used to search for excited quarks decaying to qg (Cashmore 1983).

It may well be that the $SU(2)_L \times U(1)$ description of the weak currents is correct, but that there are additional $W^{\pm}$ at higher masses, coupling in the same way. If so, the W propagator is modified, with $(Q^2+M_W^2)^{-1}$ being replaced by $K/(Q^2+M_W^2)+(1-K)/(Q^2+M_2^2)$. With $K = 0.9$, not excluded at present, the cross section for $e^- p \rightarrow \nu X$ is increased by 30% for $M_2 = 200$ GeV at $Q^2 = 16000\ \mathrm{GeV}^2$, and by 50% as $M_2 \rightarrow \infty$. Based on a run of $100\ \mathrm{pb}^{-1}$, we conclude that we are sensitive, down to 5% of the Fermi coupling, to $200 < M_2 < \infty$. This is in complete contrast to the $p\bar{p}$ collider at CERN or even to the Tevatron collider where a run of $100\ \mathrm{pb}^{-1}$ produces only two events for $M_2 = 400$ GeV.

Alternatively there may be a right-handed weak charged current, coupling to a massless or massive neutrino ($M_W > 400$ GeV for a massless neutrino (Stoker *et al.* 1985)). HERA offers the opportunity to study $e_R^+ p \rightarrow \nu X$, which is otherwise forbidden. The theoretical maximum beam helicity is 93% and helicities in the range 0.6–0.8 are expected. Pessimistically picking a helicity of 0.5, we find that in a run of $100\ \mathrm{pb}^{-1}$, the rate for this process can be measured down to 1% of the normal CC rate.

Another possibility is that the GSW theory should be extended to $SU(2)_L \times U(1) \times SU(2)$ or $SU(2)_L \times U(1) \times U(1)$. In either case there will be a new Z_2 neutral boson and a new mixing angle $\sin^2\theta_2$. For $\sin^2\theta_2 = 0.5$, present neutral-current data permit $m(Z_2)$ as low as 200 GeV. The presence of a new Z_2 will modify the cross sections for e_L^-, e_R^-, e_L^+, e_R^+ and careful measurements will reveal $m(Z_2)$ and $\sin^2\theta_2$. By measuring charge asymmetries, such as $A = [\sigma(e^-_L)-\sigma(e^+_R)]/[\sigma(e^-_L)+\sigma(e^+_R)]$, as a function of y, Z_2 masses up to 500 GeV will be accessible.

Another extension of the standard model to be considered is that there are new fermion states. At present each generation of fermions is classified into left-handed weak isodoublets and right-handed singlets:

$$\begin{pmatrix}\nu\\ e\end{pmatrix}_L, \begin{pmatrix}u\\ d\end{pmatrix}_L, e_R, u_R, d_R.$$

There may be additional states, such as

$$\begin{pmatrix}L^0\\ e\end{pmatrix}_R, \begin{pmatrix}u\\ D\end{pmatrix}_R \text{ or } \begin{pmatrix}U\\ d\end{pmatrix}_R.$$

The present data set $m_{U,D,L} > 20$ GeV (Burger 1981; Sakurai 1981; Donoghue & Holstein 1982). New Q, L can be produced by the exchange of existing or new W and Zs and spectacular final states are produced. For example, the reaction $eq \rightarrow L^0Q$ with $M_Q > M_L$ gives $e^+e^-e^-l^+l^-q$, $e^+e^-e^-lq\bar{q}$ and $e^+e^-e^-q\bar{q}q\bar{q}q$ with branching ratios of $\frac{1}{9}$, $\frac{4}{9}$ and $\frac{4}{9}$. Each quark is to be understood as a high p_T jet. Placing a limit of one event per ten days one finds one can access (M_L+M_Q) up to 220 GeV if produced by W exchange, or (M_L+M_Q) up to 175 GeV if a W_{2L} of mass 200 GeV is exchanged.

If leptons and quarks are composites with common components then new interactions are possible. We know from QED (quantum electrodynamics) tests in $e^+e^- \rightarrow e^+e^-$, for example, that the scale of any such interaction is of the order of 1 TeV. However, this by no means excludes it from study. In addition to the exchanges of γ, W, Z, we will see an effective four-fermion interaction (see figure 8*a*). This is analogous to the Fermi theory of β decay obtained by setting $|Q^2| \ll M_W^2$ and thus suppressing the W propagator. The structure function will be modified, but it is the interference of the residual interaction with the standard model that extends the sensitivity to well beyond the kinematic limit (Ruckl 1983, 1984). (Compare the measurement of electroweak asymmetries at PETRA (Althoff *et al.* 1984; Saxon 1984).) One has for the Lagrangian,

$$L_{\text{eff}} = L_{\text{SM}} \pm g^2/\Lambda^2 \sum C(\bar{e}\Gamma e)(\bar{q}\Gamma q),$$

where the Γ matrices may have various chiral structures ($g^2/4\pi$ is set equal to one conventionally for these superstrong interactions). Large polarization asymmetries result. In figure 8*b* we show an asymmetry comparing e^-_L and e^-_R which probes the lepton coupling up to $\Lambda \approx 5$ TeV. Comparing e^-_L and e^+_R will probe the quark coupling.

Supersymmetry has been proposed an explanation for the hierarchy problem, and naturally produces SUSY partners of quarks and leptons on the W and Z mass scale. The reactions with the greatest sensitivity are $eq \rightarrow \tilde{e}\tilde{q}$ by $\tilde{\gamma}$ exchange (Jones & Llewellyn-Smith 1983) and $eq \rightarrow eq$ followed by SUSY bremsstrahlung $e \rightarrow \tilde{e}\tilde{\gamma}$ (Salati & Wallet 1983). A cross section of 0.1 pb allows access to $M(\tilde{e})+M(\tilde{q})$

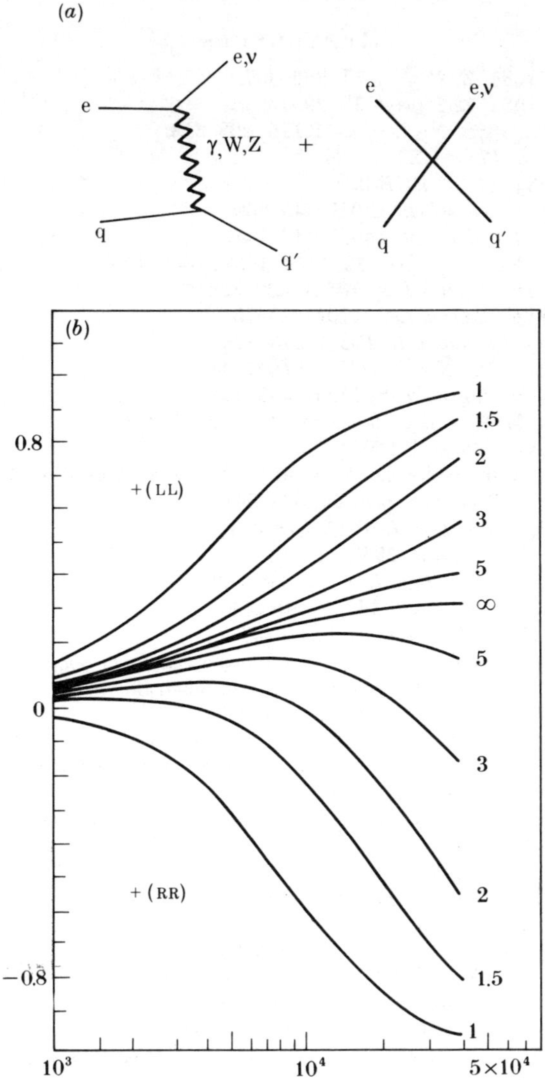

FIGURE 8. Substructure effects in eq scattering. (*a*) Effective four-fermion interaction interfering with standard processes. (*b*) The left-handed, right-handed cross-section asymmetry at $x = 0.5$ plotted against Q^2 for LL and RR interactions. Curves show effects of different scales Λ/TeV; $\Lambda = \infty$ gives the standard model.

up to 180 GeV. The event signatures are spectacular: (high-p_T lepton) + (high-p_T jet) + (missing p_T) + (y discrepancy lepton and jet). This allows great sensitivity and background suppression.

We can summarize all this briefly. The mass range of above M_W, M_Z is a fertile ground of theoretical speculation in which ep collisions at HERA will have a decisive impact. The ability to adjust the beam charge and helicity is a most powerful tool that both complements and extends the possibilities in e^+e^- collisions. Lepton–hadron scattering has been extraordinarily fruitful and instructive in the past, and shows promise of opening new physics windows in the future.

References

Abramowicz, H. *et al.* 1983 *Z. Phys.* C**17**, 283–307.
Abramowicz, H. *et al.* 1984 *Z. Phys.* C**25**, 29–43.
Altarelli, G. & Parisi, G. 1977 *Nucl. Phys.* B**126**, 298–318.
Althoff, M. *et al.* 1984 *Z. Phys.* C**22**, 13–21.
Arnold, R. H. *et al.* 1983 *SLAC-PUB*-3257.
Argento, A. *et al.* 1983 *Physics Lett.* **120**B, 245–250.
Argento, A. *et al.* 1984 *Physics Lett.* **140**B, 142–144.
Aubert, B. *et al.* 1974 *Phys. Rev. Lett.* **32**, 1454–1457; 1457–1460.
Aubert, J. J. *et al.* 1981*a* *Physics Lett.* **105**B, 322–328.
Aubert, J. J. *et al.* 1981*b* *Physics Lett.* **110**B, 73–76.
Aubert, J. J. *et al.* 1983 *Physics Lett.* **123**B, 275–278.
Aubert, J. J. *et al.* 1985*a* *CERN rep. CERN-EP*/85-34.
Aubert, J. J. *et al.* 1985*b* *Physics Lett.* **152**B, 433–438.
Bagger, J. A. & Peskin, M. E. 1985 *Phys. Rev.* D**31**, 2211–2222.
Bailey, J. *et al.* 1979 *Nucl. Phys.* B**150**, 1–75.
Bardeen, W. A., Buras, A. J., Duke, D. W. & Muta, T. 1978 *Phys. Rev.* D**18**, 3998–4017.
Barish, B. C. *et al.* 1975 *Phys. Rev. Lett.* **34**, 538–541.
Barish, S. J. *et al.* 1974 *Phys. Rev. Lett.* **33**, 448–451.
Bartel, W. *et al.* 1968 *Physics Lett.* **28**B, 148–151.
Benvenuti, A. *et al.* 1974 *Phys. Rev. Lett.* **32**, 800–803.
Bergsma, F. *et al.* 1984 *Physics Lett.* **147**B, 481–486.
Bodek, A. *et al.* 1983 *Phys. Rev. Lett.* **51**, 534–537.
Bodek, A. 1984 In *Proc. 11th int. conf. on neutrino physics and astrophysics* (ed. K. Kleinknecht & E. A. Paschos), pp. 643–669. Singapore: World Scientific.
Bolognese, T. *et al.* 1983 *Phys. Rev. Lett.* **50**, 224–227.
Bouchiat, M. A., Guena, J., Pottier, L. & Hunter, L. 1984 *Physics Lett.* **134**B, 463–468.
Brodsky, S. J., Hoyer, P., Peterson, C. & Sakai, N. 1980 *Physics Lett.* **93**B, 451–455.
Buras, A. J. 1980 *Rev. mod. Phys.* **52**, 199–276.
Bürger, J. 1981 In *Proc. int. symp. on lepton and photon interactions at high energies, Bonn* (ed. W. Pfeil), pp. 115–137.
Callan, C. G. & Gross, D. G. 1968 *Phys. Rev. Lett.* **21**, 311–313.
Callan, C. G. & Gross, D. G. 1969 *Phys. Rev. Lett.* **22**, 156–159.
Cashmore, R. J. 1983 In *HERA* 1983, pp. 381–448.
Cashmore, R. J. *et al.* 1985 *Physics Rep.* **122**, 275–386.
Chadha, S., Proudfoot, J. & Saxon, D. H. 1983 *Rutherford rep. RL*-83-071.
Clark, A. R. *et al.* 1983 *Phys. Rev. Lett.* **51**, 1826–1829.
Close, F., Roberts, R. G. & Ross, G. G. 1983 *Physics Lett.* **129**B, 346–350.
Close, F., Jaffe, R. L., Roberts, R. G. & Ross, G. G. 1984 *Phys. Rev.* D**31**, 1004–1013.
Cohen, E. R. & Taylor, B. N. 1973 *J. Phys. Chem. Ref. Data* **2**, 663–734.
Commins, E. D. & Bucksbau, P. H. 1980 *A. Rev. nucl. part. Sci.* **30**, 1–52.
Devoto, A., Duke, D. W., Owens, J. F. & Roberts, R. G. 1983 *Phys. Rev.* D**27**, 508–522.
Dokshitser, Yu. L., Dyakanov, D. I. & Troyan, S. I. 1978 *Physics Lett.* B**79**, 269–272.
Donoghue, J. F. & Holstein, B. R. 1982 *Physics Lett.* **113**B, 382–386.
Drees, J. 1981 In Bürger (1981), pp. 474–507.
Duke, D. W. & Owens, J. F. 1984 *Phys. Rev.* D**30**, 49–54.
Duke, D. W. & Roberts, R. G. 1985 *Physics Rep.* **120**, 275–368.
Dydak, F. 1983 In *Proc. int. symp. on lepton and photon interactions at high energies* (ed. D. Cassel & D. L. Kreinick), pp. 634–685. *Cornell*.
Ericson, M. & Thomas, A. 1983 *Physics Lett.* **128**B, 112–116.
Fahri, E. & Susskind, L. 1981 *Physics Rep.* **74**C, 277–321.
Feltesse, J. 1985 Rapporteur's Talk at EPS Meeting, Bari.
Fisk, H. E. & Sciulli, F. 1982 *A. Rev. nucl. part. Sci.* **32**, 499–573.
Friedman, J. I. & Kendall, H. W. 1972 *A. Rev. nucl. Sci.* **22**, 203–254.
Geweniger, C. 1984 In Bodek (1984), pp. 265–279.
Glashow, S. 1961 *Nucl. Phys.* **22**, 579–588.
Gluck, M. & Reya, E. 1978 *Physics Lett.* **79**B, 453–458.

Gluck, M. & Reya, E. 1979 *Physics Lett.* **83**B, 98–102.
Greene, G. L. *et al.* 1979 *Phys. Rev.* D**20**, 2139–2153.
de Groot, J. G. H. *et al.* 1979 *Z. Phys.* C**1**, 143–162.
Gross, D. J. & Llewellyn-Smith, C. H. 1969 *Nucl. Phys.* B**14**, 337–347.
Haidt, D. 1984 *DESY rep.* 84–108.
Hasert, F. J. *et al.* 1973*a* *Physics Lett.* **46**B, 121–124.
Hasert, F. J. *et al.* 1973*b* *Physics Lett.* **46**B, 138–140.
HERA 1983 *Proc. workshop on experimentation at HERA, Amsterdam* 1983; *DESY HERA* 83/20.
HERA 1985 *Proc. discussion meeting on HERA experiments, Genoa* 1984; *DESY HERA* 85/01.
Hofstadter, R. 1956 *Rev. mod. Phys.* **28**, 214–253.
't Hooft, G. 1971*a* *Nucl. Phys.* B**33**, 173–199.
't Hooft, G. 1971*b* *Nucl. Phys.* B**35**, 167–188.
Jones, S. K. & Llewellyn-Smith, C. H. 1983 *Nucl. Phys.* B**217**, 145–171.
Lane, K. D. & Peskin, M. E. 1980 *Proc. 15ème Rencontre de Moriond*, vol. 2 (ed. J. Tran Thanh Van), pp. 469–511.
Leveille, J. P. & Weiler, T. 1979 *Nucl. Phys.* B**147**, 147–173.
Lipkin, H. J. 1969 *Proc. Lund int. conf. on elementary particles* (ed. G. von Dardel), pp. 51–60.
Llewellyn-Smith, C. H. 1970 *CERN rep. TH*-1188.
Llewellyn-Smith, C. H. 1983 *Physics Lett.* **128**B, 107–111.
Longo, E. 1983 In *HERA* 1983, pp. 285–380.
Marciano, W. J. 1984 In *Proc. 4th topical workshop on proton–antiproton collider physics* (*Bern* 1984) (ed. H. Hänni & J. Schacher), pp. 165–180. CERN rep. 84-09.
Meyers, P. 1983 Ph.D. thesis *LBL*-17108, University of California.
Panofsky, W. K. H. 1968 *Proc. 14th int. conf. on high-energy physics, Vienna* 1968 (ed. J. Prentki & J. Steinberger), pp. 23–39. CERN.
Particle Data Group 1984 *Rev. mod. Phys.* **56**, no. 2, part 2.
Perkins, D. H. 1972 In *Proc. 16th int. conf. on high-energy physics, Batavia*, vol. 4 (ed. J. D. Jackson & A. Roberts), pp. 189–248.
Perkins, D. H. 1977 *Rep. Prog. Phys.* **40**, 409–481.
Perkins, D. H. 1982 *Introduction to high-energy physics* (2nd edn). London: Addison Wesley.
Perkins, D. H. 1983 In *HERA* 1983, pp. 39–54.
Prescott, C. *et al.* 1978 *Physics Lett.* **77**B, 347–352.
Prescott, C. *et al.* 1979 *Physics Lett.* **84**B, 524–528.
Riordan, E. M. *et al.* 1975 *SLAC-PUB*-1634.
Ruckl, R. 1983 *Physics Lett.* **129**B, 363–369.
Ruckl, R. 1984 *Nucl. Phys.* B**234**, 91–105.
Sakurai, J. J. 1981 In *Proc. int. conf. on neutrino physics and astrophysics, Maui, Hawaii*, vol. 2 (ed. R. J. Cence, E. Ma & A. Roberts), pp. 457–491.
Salam, A. 1968 In *Elementary particle theory, relativistic groups and analyticity* (Nobel Symposium no. 8) (ed. N. Svartholm), pp. 367–377. Stockholm: Almqvist & Wiksell.
Salati, P. & Walet, J. C. 1983 *Physics Lett.* **122**B, 397–402.
Sarkar-Cooper, A. M. 1984 In Bodek (1984), pp. 394–404.
Savin, I. 1984 In *Proc. 22nd int. conf. on high-energy physics, Leipzig* 1984, vol. 2 (ed. A. Meyer & E. Wieczorek), pp. 251–278.
Saxon, D. H. 1984 *Electroweak interference in* e^+e^- *annihilation. Proc. Phys. in Collision 4* (ed. A. Seiden), pp. 295–316. Gif sur Yvette: Editions Frontières.
Schrempp, B. & Schrempp, F. 1985 *Physics Lett.* **153**B, 101–107.
Schwinberg, P. B., Van Dyck, R. S. & Dehmelt, H. G. 1981 *Phys. Rev. Lett.* **47**, 1679–1682.
Segrè, E. 1980 *From X-rays to quarks.* San Francisco: Freeman & Co.
Söding, P. & Wolf, G. 1981 *A. Rev. Nucl. Part. Sci.* **31**, 231–293.
Stoker, D. P. *et al.* 1985 *Phys. Rev. Lett.* **54**, 1887–1890.
Weinberg, S. 1967 *Phys. Rev. Lett.* **19**, 1264–1266.
Winter, K. 1984 *CERN rep., CERN-EP*/84-137.
Wotschak, J. 1981 In Bürger (1981), pp. 461–473.
Wu, S. L. 1984 *Physics Rep.* **107**, 59–324.
Yukawa, H. 1935 *Proc. Phys. Math. Soc. Japan*, **17**, 48.